ESTIMATING FOR CIVIL ENGINEERS

Dr. D.V. Varshney
B.Sc., B.Sc. (Engg.) (Hons.)
P.G. Dip., M.E., Ph.D. (Engg.), F.I.E.
Professor in Civil Engineering
G.B. Pant University, Pantnagar - 263 145

CBS

CBS PUBLISHERS & DISTRIBUTORS PVT. LTD.

New Delhi • Bengaluru • Chennai • Kochi • Mumbai • Pune

Dedication

Dedicated to the memory of my revered mother,
Smt. Pancho Devi, who is no more but
whose memory only remains.

D.V.

ISBN: 81-239-0335-9

First Edition: 1994
Reprint: 2011, 2012, 2014, 2015

Published by:
Satish Kumar Jain for CBS Publishers & Distributors Pvt. Ltd.,
4819/XI Prahlad Street, 24 Ansari Road, Daryaganj, New Delhi - 110002
delhi@cbspd.com, cbspubs@airtelmail.in • www.cbspd.com
Ph.: 23289259, 23266861, 23266867 • Fax: 011-23243014

Corporate Office: 204 FIE, Industrial Area, Patparganj, Delhi - 110 092
Ph: 49344934 • Fax: 011-49344935
E-mail: publishing@cbspd.com • publicity@cbspd.com

Branches:

- ***Bengaluru:*** 2975, 17th Cross, K.R. Road, Bansankari 2nd Stage, Bengaluru - 70 • Ph: +91-80-26771678/79 • Fax: +91-80-26771680 E-mail: cbsbng@gmail.com, bangalore@cbspd.com
- ***Chennai:*** No. 7, Subbaraya Street, Shenoy Nagar, Chennai - 600030 Ph: +91-44-26681266, 26680620 • Fax: +91-44-42032115 E-mail: chennai@cbspd.com
- ***Kochi:*** 36/14, Kalluvilakam, Lissie Hospital Road, Kochi - 682018 Ph: +91-484-4059061-65 • Fax: +91-484-4059065 E-mail: cochin@cbspd.com
- ***Mumbai:*** 83-C, Dr. E. Moses Road, Worli, Mumbai - 400018 Ph: +91-9833017933, 022-24902340/41 • E-mail: mumbai@cbspd.com
- ***Pune:*** Bhuruk Prestige, Sr. No. 52/12/2+1+3/2, Narhe, Haveli (Near Katraj-Dehu Road Bypass), Pune - 411041 Ph: +91-20-64704058/59, 32342277 • E-mail: pune@cbspd.com

Representatives:

- Hyderabad: 0-9885175004
- Nagpur: 0-9021734563
- Vijayawada: 0-9000660880
- Kolkata: 0-9831437309, 0-9051152362
- Patna: 0-9334159340

Printed at:
J.S. Offset Printers, Delhi

PREFACE

Civil engineers are required to give realistic estimates of works at planning and execution stage. The correctness and realistic estimation forms a very important task for an engineer. An effort has been made in this book to familiarise the students with the techniques of performing analysis of rates and estimation. Illustrative examples of various works have been given to guide the students. The engineers in the field will find the book very helpful in their day-to-day work. Any suggestion for the improvement of the book are most welcome.

D.V. Varshney

CONTENTS

1

ANALYSIS OF RATES

Rate of a particular item is determined after taking into account various factors affecting the cost of item. This process of determining the rate of an item is termed as analysis of rate. The process of rate analysis gives a clear picture of various components and operations involved in the performance or manufacture of the said item.

The factors which affect the rate analysis of an item can broadly be divided into two categories (a) Major factors; (b) Minor factors.

Major Factors

There are mainly three factors on which the rate of an item depends, namely (i) Material (ii) Labour and (iii) Operational cost of equipments.

(i) *Material:* The quantities of various materials required for the construction of an item, depends upon the specifications of the item. The prices of various materials will depend upon the market conditions; varying from place to place; and from time to time. Hence, before starting the rate analysis of an item it is essential to collect the prices of such materials from the market at that instant. In Government departments, normally the schedule of rate covers the rates of materials in a particular locality. With the help of the quantities of various materials and prices of the materials from the market or schedule of rate the cost of materials for a particular item can be calculated. In order to arrive at the cost of material at the site of work, the cost of transport from the place of availability to site has to be accounted for as per the schedule of rates for transport.

(ii) *Labour:* The amount of labour force of different categories that may be required to carry out a particular operation of the item of work is estimated on the basis of experience and standard norms laid down by any standardizing agency. There may be certain variations from the norms due to the climatic or other local conditions, for which the Engineer-in-charge is supposed to provide ample justifications in his analysis of rates. In case of certain items no guidelines may be available and so the Engineer-in-charge may carry on his own sample work and prepare the analysis on the basis of that.

(iii) *Operational Cost of Equipments:* Sometimes the execution of the particular item of work may involve use of special equipment, machinery and tools and plants. In such a case the operational cost of the machinery should be added per unit quantity of the item in question.

Overhead Costs: The overhead costs include establishment charges, stationery charges, postage, printing, travelling expenses, telephone, rents, taxes, royalties, interests on investments and losses, etc.

Normally a contractor's profit of 10% may be permissible according to the norms. For overhead expenses and contractor's profit 15% of the actual cost may be reasonably included in the analysis of rates.

In order to facilitate preparation of the analysis of rates, the norms of labour and material components, as adopted in the departments (Irrigation and P.W.D.) are given as below.

Lime Concrete Normally for 100 m^3 lime concrete, 100 m^3 ballast and 50 m^3 of fine mortar is required.

S. No.	Item	Mix	Unit	White lime	Surkhi	Kankar lime	Sand	Ballast
1.	Lime, Sand Concrete (for 1 m^3)	1 : 1	cu. m	0.285	–	–	0.285	1
2.	Lime, Surkhi, Sand Concrete	1 : 1 : 1	,,	0.19	0.19	–	0.19	1
3.	Lime, Surkhi, Sand Concrete	1 : 1 : 2	,,	0.14	0.14	–	0.29	1
4.	Lime, Surkhi Concrete	1 : 2	,,	0.19	0.38	–	–	1
5.	Lime, Sand Concrete	1 : 2	,,	0.19	–	–	0.38	1
6.	Lime, Sand Concrete	1 : 3	,,	0.14	–	–	0.43	1
7.	Lime Ash, Sand Concrete	1 : 1 : 1	,,	0.19	0.19 (Ash)	–	0.19	1
8.	Kankar Lime Concrete	–	–	–	–	0.40	–	1

Cement Concrete

S. No.	Mixes	Unit by Volume	Natural Aggregate			Crushed Aggregate			Remarks
			Cement in bags of 50 kg	Sand in cu. m	Coarse aggregate in cu. m	Cement in bags	Sand in cu. m	Coarse aggregate in cu. m	
1.	1 : 1 : 2	1 cu. m	10.8	0.30	0.75	11.10	0.39	0.78	
2.	1 : 1.5 : 3	,,	7.56	0.40	0.80	7.90	0.41	0.83	
3.	1 : 2 : 3	,,	7.20	0.50	0.75	7.56	0.52	0.78	
4.	1 : 2 : 4	,,	5.76	0.40	0.80	6.10	0.42	0.85	
5.	1 : 2.5 : 5	,,	4.95	0.43	0.86	5.20	0.45	0.90	
6.	1 : 3 : 6	,,	4.15	0.43	0.86	4.30	0.45	0.90	
7.	1 : 4 : 8	,,	3.25	0.45	0.90	3.40	0.46	0.93	
8.	1 : 5 : 10	,,	2.50	0.44	0.99	2.70	0.47	0.94	
9.	1 : 6 : 12	,,	2.15	0.45	0.90	2.35	0.49	0.98	
10.	1 : 8 : 16	,,	1.70	0.47	0.94	1.80	0.50	1.00	

The ingredients mentioned above are based on usual conventional practice, and the size of the aggregate has not been taken into account. The size and grading of the aggregate have an important bearing on the water requirement, and consequently cement content. It is recommended that on important works, involving large quantities of concrete, the concrete mixes should be designed, specifying the strength and size of aggregate on the basis of laboratory tests.

Masonry Work

S. No.	Mix of Mortar	Unit	Cement in bags of 50 kg	Sand in cu. m	Number of bricks
1.	1 : 3	Cu. m of brickwork	2.60	0.275	460
2.	1 : 4		2.00	0.275	460
3.	1 : 5		1.60	0.275	460
4.	1 : 6		1.35	0.275	460

Size of brick in the above has been taken as 23 cm × 11 cm × 7 cm, with 9.5 mm thick joint in between.

If the size of bricks or thickness of joint is different, the sand and brick requirements may be worked out as follows.

Sand in cu. m $M = \left(\dfrac{ST/V + 0.1}{ST/V + 2}\right)$

Cement in bags of 50 kg $= \dfrac{M}{R} \times 28.75$

where,

S = Total Surface area of one brick

V = Total Volume of one brick, neglecting frog

T = Thickness of joint

R = Ratio of mortar mix (1 : R)

Stone Masonry in Cement Mortar

S. No.	Name of Item	Unit	Mortar Required in m^3	Cement in Bags	Sand in cu. m	Stone in cu. m
1.	Random Rubble 1 : 4	cu. m	0.45	3.2	0.45	1
2.	Random Rubble 1 : 3	"	0.45	4.3	0.45	1
3.	Coursed Stone (Dressed) 1 : 3	"	0.33	3.15	0.33	1.05

Cement Pointing

S. No.	Mix	Unit	Cement Bags (of 50 kg)	Sand in cu. m
1.	1 : 2	sq. m	0.055	0.0036
2.	1 : 3	sq. m	0.035	0.0036

Note : For stone masonry, the quantities are increased by 33%.

Cement Plaster

S. No.	Mix	Unit	Cement Bags of 50 kg	Sand in cu. m
1.	1 : 3	1.25 cm thick cement plaster per sq. m of surface area.	0.14	0.015
2.	1 : 4		0.11	0.015
3.	1 : 5		0.095	0.0165
4.	1 : 6		0.079	0.0165

Note : Quantities are increased by 33% for stone masonry.

Reinforced Brick Work

Reinforced Brick Work in 1 : 3 Cement Mortar

Mortar — Quantities of Ingredients per cubic metre

(i) Cement — 3.5 bags of 50 kg each

(ii) Sand — 0.36 cubic metre

(iii) Bricks — 23 cm × 11 cm × 7 cm — 405 numbers

Note :

(a) A minimum cover of 6 mm on either side of reinforcing bars shall be provided. Thickness of vertical joints shall be as follows:

for 6 to 8 mm dia. bars	20 mm
9 to 11 mm dia. bars	23 mm
12 to 14 mm dia. bars	26 mm
16 mm	28 mm

(b) Thickness of horizontal joints is 10 mm minimum, if the size of bricks is different. The quantities of ingredients per unit of brick work in 1 : 3 can be computed by the following formula.

(i) Number of bricks $\dfrac{1.04}{(L+T)(D+T)\times B}$

(ii) Sand $1.05\left[\dfrac{BT(L+T+D)+0.4\,L\,.\,E\,.\,T}{(L+T)(D+T)\times B}\right]$

(iii) Cement $\dfrac{\text{Sand}}{3}$

where *L*, *B* and *D* denote the length, breadth and depth of bricks.

T = Thickness of joints.

Carriage of Materials

S. No.	Item	Details	Remark
1.	Time of loading and unloading		
	Loading	15 mts	
	Unloading	10 mts	
	Miscellaneous	5 mts	
	Total	30 mts	
2.	Time of carriage one km (both sides)	30 mts	
3.	Cost of constants/day cart including		
	Cartman @ Rs *A*/day	Rs *A* per day	
	One labour @ Rs *B*/day	Rs *B* per day	
	Total	Rs (*A* + *B*)/day	
4.	Contractor's profit	12.5%	
5.	Number of trips/day (8 hours)	$N=\dfrac{8}{\frac{30}{60}\,.\,L+\frac{1}{2}}$	*L* = lead in km

Rate of carriage per bullock cart — *R*

Rate of carriage per bullock cart per trip for lead of *L* kms

$$R = 1.125\,\frac{(A+B)}{N}$$

$$= 1.125\,\frac{(A+B)}{8}\times\left(\frac{1}{2}L+\frac{1}{2}\right)$$

$$= 0.07\,(A-B)\,(L+1)$$

where

A = Hiring charges of cart including cartman/day
B = Daily labour rate of one labour
L = Lead in km one side

Carrying Capacity of One Bullock Cart

S. No.	Material	Quantity of Material	
		by Pucca Road	by Kacha Road
1.	Bricks — 23 × 11 × 7 cms	250	225
2.	Coal	7.5 Quintals	6.75 Quintals.
3.	Steel	7.5 Quintals	6.75 Quintals
4.	Cement	15 bags (50 kg each)	14 bags (50 kg each)
5.	Sand	0.45 cu. m	0.41 cu. m
6.	Brick Rora/Brick ballast/Kankar	0.70 cu. m	0.63 cu. m
7.	Stone ballast/Stone	0.51 cu. m	0.46 cu. m
8.	Lime/Surkhi	0.625 cu. m	0.56 cu. m

Note : 1. If the size of bricks is different, the carrying capacity may be modified in the ratio of volumes.

2. Transport of materials by bullock carts should be resorted to only where power transport is either not available or not permissible.

Capacity of Labour

S. No.	Item of Work	Units	Capacity of one Beldar per day	Labour required	Rates			Including, 12.5% contractors profit
					Beldar Rs *A*/day	Mason as Rs *B*/day	Mate as Rs *C*/day	
1.	Dry Earth Work in foundation	cum	2.8 cum	Beldar 1/2.8 Mason 1/28	0.358 *A*	0.0358 *B*	–	0.04 (10 *A* + *B*)
2.	Wet Earth Work in foundation							
	(a) Wet With Water	cum	2.0 cum	Beldar 1/2 Mason 1/28	0.5 *A*	0.0358 *B*	–	0.04 (14 *A* + *B*)
	(b) Wet without water	cum	2.5 cum	Beldar 2/5 Mason 1/28	0.4 *A*	0.0358 *B*	–	0.04 (11 *A* + *B*)
3.	Earth Work in canals and branches	cum	2.8 cum	Beldar 1/2.8 Mate 1/112	0.358 *A*	–	0.09 *C*	0.04 (10 *A* + 0.25 *C*)
4.	Earth Work in Distributaries	cum	3.1 cum	Beldar 1/3.1 Mate 1/112	0.322 *A*	–	0.09 *C*	0.04 (9 *A* + 0.25 *C*)
5.	Earth Work in Drains and Minors	cum	3.5 cum	Beldar 2/7 Mate 1/140	0.286 *A*	–	0.007 *C*	0.04 (8 *A* + 0.2 *C*)
6.	Earth Work in Guls	cum	4.0 cum	Beldar 1/4 Mate 1/140	0.25 *A*	–	0.007 *C*	0.04 (7 *A* + 0.2 *C*)
7.	Silt clearance in distributaries	cum	3.5 cum	Beldar 2/7 Mate 1/140	0.286 *A*	–	0.007 *C*	0.04 (8 *A* + 0.2 *C*)
8.	Silt clearance in Drains and Minors	cum	4.0 cum	Beldar 1/4 Mate 1/140	0.25 *A*	–	0.007 *C*	0.04 (7 *A* + 0.02 *C*)
9.	Earth Work in Embankments over 1.5 m high	cum	2.25 cum	Beldar 4/9 Mate 1/56	0.44 *A*	–	0.178 *C*	0.04 (12.5 *A* + 0.5 *C*)
10.	Extra for compaction of high embankments	cum	2.8 cum	Beldar 1/28	0.0358 *A*	–	–	0.04 *A*
11.	Extra for lead beyond 30 m lead upto a total lead of 300 m	cum	2.8 cum	Beldar 1/28	0.0358 *A*	–	–	0.04 *A*
12.	Extra for lead beyond 200 m upto total lead of 600 m for 100 m or part there of by donkeys/mules	cum	–	Equivalent to 1/7 Beldar	0.143 *A*	–	–	0.16 *A*
13.	Extra for left above 1.5 m, for every 0.3 m or part thereof	cum	85 cum	Beldar 1/25	0.0118 *A*	–	–	0.0132 *A*
14.	Extra for mud, kabar and kankar soil	cum	19 cum	Beldar 1/19	0.053 *A*	–	–	0.06 *A*
15.	Top dressing of earth work in channels and drains	M. km	1/5 M. km	Beldar 5 Mate 1/5	0.5 *A*	–	0.2 *C*	2.25 (2.5 *A* + 0.1 *C*)

Continued

S. No.	Item of Work	Units	Capacity of one Beldar per day	Labour required	Rates			Including 12.5% contractors profit
					Beldar Rs *A*/day	Mason as Rs *B*/day	Mate as Rs *C*/day	
16.	Levelling and dressing of service road	M. km	1/5 M. km	Beldar 5 Mate 1/5	0.5 *A*	–	0.2 *C*	2.25 (2.5 *A* + 0.1 *C*)
17.	Dag belling 15 cm deep	Line/km	2/5 km	Beldars 2.5 Mate 1/16	2.5 *A*	–	0.0625 *C*	0.7 (4 *A* + 0.1 *C*)
18.	Dag belling 7.5 cm deep	Line/km	4/5 km	Beldars 11/4 Mate 1/32	1.25 *A*	–	0.032 *C*	0.7 (2 *A* + 0.05 *C*)

Note : 1. The rates included upto 30 m lead and 1.5 m lift tough dressing and normal compaction.

2. Earth work in repairs should be measured at borrow pits and in embankment by the sectional area of the finished work.

Capacity of Labour

S. No.	Item of Work	Unit	Capacity of mason per day	Labour Requirement per day: Mason (A)	Labour (B)	Bhisti (C)	Mistri (D)	Total rate including 12.5% contractor's profit	Remarks
1. Masonry									
1.	Brick work in mud upto 5 m height with 60 m lead of material	cum	1.41 cum	0.71	1.42	0.088	0.088	$\left(2A+4B+\frac{1}{4}C+\frac{1}{4}D\right)\times 0.4$	Wages per day Mason - Rs *A* Labour - Rs *B* Bhisti - Rs *C* Mistri - Rs *D* Carpenter - Rs *E* Blacksmith - Rs *F* Rent of mixing Rs *L* It does not include dressing of stone.
2.	Brick work in cement mortar upto 3 m height with 60 m lead of materials	cum	1.31 cum	0.88	1.76	0.176	0.088	$\left(2.5A+5B+\frac{1}{2}C+\frac{1}{4}D\right)\times 0.4$	
3.	Uncoarsed Rubble or Random Rubble Masonry upto 3 m height with 60 m lead of materials	cum	1.31 cum	0.88	1.76	0.176	0.088	$\left(2.5A+5B+\frac{1}{2}C+\frac{1}{4}D\right)\times 0.4$	
4.	Coursed Rubble Masonry upto 3 m height with 60 m lead of materials	cum	0.79 cum	1.24	1.76	0.176	0.088	$\left(3.5A+5B+\frac{1}{2}C+\frac{1}{4}D\right)\times 0.4$	
5.	Extra for every 1.5 m above 3 m height including scaffolding	cum	–	–	0.35	–	–	$(B)\times 0.4$	
6.	Extra for lead of 30 m, beyond 60 m	cum	–	–	0.176	–	–	$\left(\frac{1}{2}B\right)\times 0.4$	
2. Arch Work									
1.	Brick Arch Work	cum	0.79	1.24	2.12	0.176	0.088	$\left(3.5A+6B+\frac{1}{2}C+\frac{1}{4}D\right)\times 0.4$	
2.	Stone Arch Work	cum	0.56	1.76	2.12	0.176	0.088	$\left(5A+6B+\frac{1}{2}C+\frac{1}{4}D\right)\times 0.4$	
3.	Centering and shuttering for Arch work	cum	–	53 carpenters	1.76	–	0.176	$\left(5B+\frac{1}{2}D+\frac{1}{2}E\right)\times 0.04$	

Continued

S. No.	Item of Work	Unit	Capacity of mason per day	Labour Requirement per day				Total rate including 12.5% contractor's profit	Remarks
				Mason (*A*)	Labour (*B*)	Bhisti (*C*)	Mistri (*D*)		
3.	**Cement Pointing and Plastering**								
1.	Cement Plaster on brick masonry	cum	2.3	0.102	0.216	0.054	–	$(A+2B+C)\times 0.12$	
2.	Cement plaster on stone masonry	sq. m	9.3 sq. m	0.108	0.216	0.05	–	$\left(A+2B+\frac{1}{2}C\right)\times 0.12$	
3.	Cement pointing (Flush) on brick masonry	,,	9.35 sq. m	0.108	0.108	0.027	–	$\left(A+B+\frac{1}{4}C\right)\times 0.12$	
4.	Cement pointing (Struck deep) on brick masonry	,,	6.15 sq. m	0.162	0.108	0.027	–	$\left(1\frac{1}{2}A+B+\frac{1}{4}C\right)\times 0.12$	
5.	Cement pointing (Flush) on stone	,,	9.3 sq. m	0.108	0.216	0.027	–	$\left(A+2B+\frac{1}{4}C\right)\times 0.12$	
6.	Cement pointing on stone (struck deep or raised)	,,	6.1 sq. m	0.162	0.216	0.027	–	$\left(1\frac{1}{2}A+2B+\frac{1}{4}C\right)\times 0.12$	
7.	Cement plaster or pointing extra for every 15 m, above 3 m height, including scaffolding	,,	–	–	0.027	–	–	$\left(\frac{1}{4}B\right)\times 0.12$	
4.	**Flooring, Pitching and Cement Concrete in Floors and Roofs**								
1.	Flat brick flooring (dry)	sq. m	186 sq. m	0.054	0.054	–	–	$\frac{1}{2}(A+B)\times 0.12$	
2.	Brick floor on edge (dry)	,,	12.3 sq. m	0.081	0.081	–	–	$\frac{3}{4}(A+B)\times 0.12$	
3.	Dry brick or stone pitching 3 m height, 30 m lead of materials	cum	1.9 cum	0.162	0.216	–	–	$\left(1\frac{1}{2}A+2B\right)\times 0.4$	

Continued

S. No.	Item of Work	Unit	Capacity of mason per day	Labour Requirement per day				Total rate including 12.5% contractor's profit	Remarks
				Mason (*A*)	Labour (*B*)	Bhisti (*C*)	Mistri (*D*)		
4.	Cement concrete flooring overbase concrete upto 4 cm.	sq. m	3.1 sq. m	0.324	0.54	0.054	0.027	$\left(3A+5B+\frac{1}{2}C+\frac{1}{4}D\right)\times 0.12$	
5.	Cement concrete or lime concrete in foundation	cum	2.83 cum	0.35	1.42	0.35	–	$(A+B+C)\times 0.4$	
6.	Cement or lime concrete on roof	,,	2.83	0.35	3.88	0.35	0.088	$\left(A+11B+C+\frac{1}{4}D\right)\times 0.4$	
7.	Extra for lead of 30 m or 1.5 m lift	,,	–	–	0.176	–	–	$\left(\frac{1}{2}B\right)\times 0.4$	
5.	**R.B. & R.C.C. Work**								
1.	Bending, building and cutting reinforcement	Qtl.	–	1.5 black-smith (*F*)	1.0	–	–	$\left(\frac{3}{4}F+B\right)\times 2.25$	
2.	R.B. Work excluding reinforcing steel and centering and shuttering	cum	0.94 cum	1.06	2.12	0.35	0.088	$\left(3A+6B+C+\frac{1}{4}D\right)\times 0.4$	
3.	Centering for R.B. Work	sq. m	–	3.24 car-penter (*E*)	1.4	–	0.54	$\left(3E+13B+\frac{1}{2}D\right)\times 0.12$	
4.	R.C.C. Work excluding reinforcing steel centering and shuttering (manual labour)	cum	1.41 cum	0.71	2.82	0.71	0.088	$\left(2A+8B+2C+\frac{1}{4}D\right)\times 0.4$	
5.	R.C.C. work excluding reinforcing steel centering and shuttering (Mech. Mixing)	,,	1.41 cum	0.71 + Rs *G*/day for mech. equipment	2.82	0.71	0.088	$\left(2A+3B+2C+\frac{1}{4}D+G\right)\times 0.4$	
6.	Centering and shuttering for R.C.C. Work	sq. m	–	4.3 carpenters @ Rs *E*/perday	2.82	–	0.081	$\left(4E+17B+\frac{3}{4}D\right)\times 0.12$	

Carriage of Materials by Truck

S. No.	Items	Basic assumptions for 5 tonne truck	Sample calculations for 5 tonne truck Diesel	Rate of carriage of 1 tonne material by truck	Remark
1.	Cost of Variables			Carriage of 1 tonne of material by truck (both included) on tarred or concrete roads for a lead of L km	
	(A) (i) Cost of truck exclusive of tyres	Rs A	2,80,000.00 Rs		
	(ii) Repair charges @ 60% over life time	Rs 0.6 A	1,68,000.00 Rs		
	(iii) Life of truck	2,85,000 kms	2,85,000 km	$= \text{Rs } 0.0227 \left[1.7\, S\, L \times \left\{ \frac{1.6\,A}{12000 \times S} + \frac{1.2\,B}{24000} + \frac{C}{3.5} + \frac{2\,D}{105} \right\} + \frac{K}{N} \right]$	
	(iv) Depreciation and repair charges of truck per km	Rs $\frac{1.6\,A}{2,85,000}$ per km	$\frac{4,448,000.00}{2,85,000.00}$ = Rs 1.51/km		
	(B) (i) Cost of 6 tyres and tubes	Rs B	6,000 × 6 = 36,000 Rs		
	(ii) Repair and casualties	Rs 0.2 B	7,200 Rs		
	(iii) Life of Tyres	24,000 kms	24,000 km		
	(iv) Hence depreciation of tyres and tubes	Rs $\frac{1.2\,B}{24,000}$ per km	$\frac{43,200}{24,000}$ = Rs 1.8/km	where, L = Lead in km (one side)	
	(C) Cost of operation materials per km			The cost of variables in the empty return journey has been assumed as 75% of that in loaded journey.	
	(i) Cost of fuel per litre	Rs C	Rs 4.00 per litre		
	Distance covered in 1 lit.	3.5 km	3.5 km	S = Average speed in km per hour, which shall be taken upto 5 km lead as 16 km/hr. Upto 16 km lead as 24 km/hr. Upto 32 km lead as 32 km/hr. Above 32 km lead as 40 km/hr.	
	Cost of fuel per km	$\frac{C}{3.5}$ per km	$\frac{4}{3.5}$ = Rs 1.14/km		
	(ii) Cost of mobil oil per lit.	Rs D	20 Rs/litre		
	Distance covered in 1 lit.	105 km	105 km		
	Cost of mobil oil per km	Rs $\frac{D}{105}$ per km	$\frac{20}{105}$ = Rs 0.19/km	A = Cost of truck in Rupees	
	(iii) Other oils and stores per km	Rs $\frac{D}{105}$ per km	Rs 0.19/km	B = Cost of tyres and tubes in Rs	
	Total cost of operation materials per km	Rs $\left(\frac{C}{3.5} + \frac{2\,D}{105} \right)$	Rs 1.52/km	C = Cost of diesel per litre in Rs $K = 1.25\,(E+F) + 6\,G + \frac{H + I + J + 0.0225\,(A+B)}{274}$	
2.	Cost of constants per day of 8 hours				
	(A) (i) Cost of staff per day of 8 working hours	Driver @ Rs E per day Cleaner @ Rs F per day	Driver's pay Rs 45/day Cleaner's pay Rs 20/per day	E = Daily labour of driver in Rs F = Daily labour of cleaner in Rs	
	(ii) Add 25% of the above for idle days	Rs $(E+F) \times \frac{1}{4}$ per day	Rs $65 \times \frac{1}{4}$ = Rs 16.25 per day	G = Daily labour rate of labourers in Rs	

Continued

S. No.	Items	Basic assumptions for 5 tonne truck	Sample calculations for 5 tonne truck Diesel	Rate of carriage of 1 tonne material by truck	Remark
	(B) Cost of labour per day of 8 working hours for loading and unloading			H = Annual road tax. I = Annual comprehensive insurance. J = Annual driver's licence fee. N = Number of trips of B working hours. When $N = \frac{8}{\frac{22}{5}+1}$ Quantities of materials equivalent to 1 tonne. Bricks (23 × 11 × 7 cms) — 333 Nos. Cement — 20 bags of 50 kg each. Sand — 0.64 cum. Brick Rora — 0.94 cum. Brick Ballast — 0.94 cum. Stone Ballast — 0.71 cum. Kankar — 0.94 cum. Kankar lime — 0.85 cum.	
	(i) Labour required	6 Nos	6 Nos		
	(ii) Rate of labour per day	Rs 6 per day	Rs 15 per day		
	(iii) Cost of labour per day	Rs 66 per day	Rs 90 per day		
	(C) Cost on account of taxes and insurance				
	(i) Road tax (annual)	Rs H per year	Rs 3,000 per year		
	(ii) Comprehensive insurance (annual)	Rs I per year	Rs 3,000 per year		
	(iii) Driver's licence fee	Rs J per year	Rs 50 per year		
	(iv) Interest on HAL initial cost @ 10% per annum (assuming scrap value)	Rs $\frac{(A+B)\times 10}{100}$ per year	$\frac{(2,80,000+36,000)\times 10}{100}$ = Rs 31,600		
	Total annual expenses (i) to (iv)	Rs $(H+I+J) + \frac{(A+B)\times 10}{100}$	Rs 37,650.00		
	Cost on account of (C) per day assuming 25% idle days	Rs $\frac{(H+I+J)}{0.75\times 365} + \frac{\frac{(A+B)\times 4.5}{2\times 100}}{0.75\times 365}$ per day	$\frac{37,650.00}{0.75\times 365}$ = Rs 137.50 per day		
	Total cost of constants per day	Rs $\frac{5}{4}(E+F)+66 + \left[\frac{H+I+J+0.0225\,(A+B)}{274}\right]$ = Rs 'K' per day (say)	$\frac{5}{4}\times 65+90+[0.0225 \times 3,16,000+6050]\times\frac{1}{274}$ = 81.25 + 90 + 48.03 = 219.28 Rs 219.28 per day		
3.	Number of trips per day	$N=\frac{B}{\frac{2L}{5}+1}$ per day where 'N' = Number of trips per day of 8 working hours	For S = 24 km/hour L = 10 km $N=\frac{8}{\frac{2\times 10}{24}}$ = 4.36 per day		

Continued

S. No.	Items	Basic assumptions for 5 tonne truck	Sample calculations for 5 tonne truck Diesel	Rate of carriage of 1 tonne material by truck	Remark
		L = Lead in kms (one side) S = Speed of Truck in kms/hour I = Loading and unloading time 1 hour			
4.	Overhead charges 1% Contractor's profit 12½%	13.5%	13.5%		

Note: The above rates are for tarred or concrete roads in plains. The rates shall be increased for other types of roads by the following percentages:

(A) Metalled roads other than tarred or concrete roads including gravel, Kankar or unmetalled canal roads — 15%

(B) Kachcha roads — 30%

(C) On hill roads, the rates shall be 30% extra over the rates for the plains for the corresponding type of roads.

(D) The above rates include journey both ways, assuming empty return journey, if the return journey is not empty, the above rates shall be reduced by 33⅓%.

2

ANALYSIS OF RATES OF SOME OF THE MACHINES

The analysis of rates of some of the machinery, usually deployed for construction work, is given as below :

1. Working rate of 0.91 cu. m Poclain Excavator

Capital cost — Rs 40.00 lacs

Life — 10,000 hours

Depreciation — $\frac{0.95 \times 40,00,000}{10,000}$ = Rs 380.00 per hour

Repairs provision — $\frac{1.50 \times 40,00,000}{10,000}$ = Rs 600.00 per hour

Work Charged Establishment

1 No. of operator @ 1200/– p.m.	=	Rs 1,200.00
1 No. of helper @ 800/– p.m.	=	Rs 800.00
½ No. of mechanic @ 1200/– p.m.	=	Rs 600.00
¼ No. of foreman @ 1600/– p.m.	=	Rs 400.00
¼ No. of chowkidar @ 800/– p.m.	=	Rs 200.00
Total		Rs 3,200.00
Add 25% for bonus, leave reserve, gratuity, etc.		Rs 800.00
		Rs 4,000.00
So, establishment per year		48,000.00
Expenditure on establishment per hour	=	$\frac{48,000.00}{1,000}$
Assuming 1,000 working hours per year	=	Rs 48.00 per hour

P.O.L. Charges

15 lit. H.S.D. @ 4.50 per litre	=	Rs 67.50 per hr
Add for lubricants, grease, etc., @ 33%	=	Rs 22.50 per hr
Add for sundries, cotton waste, etc., @ 10%	=	Rs 6.75 per hr
Total		Rs 96.95 per hr say Rs 97.00

So, total expenditure per hour of operation

Depreciation	Rs 380.00
Repairs	Rs 600.00
W.C. Estt.	Rs 48.00
P.O.L.	Rs 97.00
Total	Rs 1,125.00 per hour

2. Working Rates of 4.0 m^3 Capacity Tippers

Capital cost = Rs 5.00 lacs

Life = 8,000 hours

Tyre cost = 0.5 lacs

Tyre life = 1500 hours

Depreciation of Tipper $= \dfrac{0.90 \times 5{,}00{,}000}{8{,}000} =$ Rs 56.25/hour

Tyre $= \dfrac{50{,}000}{1{,}500} =$ Rs 33.33/hour

Running Repair of Tripper $= 1.40\left(\dfrac{5{,}00{,}000 - 50{,}000}{8{,}000}\right) =$ Rs 78.75

Tyre = 15% of its depreciation = 4.9995 say = 5.00 Rs/hour

W.C. Estt.

1. 1 No. operator @ 1,500/– per 100 hours	=	1,500.00
2. 1 No. helper @ 900/– per 100 hours	=	900.00
3. ¼ No. mechanic @ 1,500/– per 100 hours	=	375.00
4. ¼ No. chowkidar @ 800/– per 100 hours	=	200.00
		2975.00
Add 25% due to L.R., bonus, gratuity, etc.		744.00
Total		3,719.00

$\therefore$ Expenses/hour $= \dfrac{3{,}719}{1000} =$ Rs 37.19/hour

Assuming 1000 working hour in one year.

P.O.L.'s

8 lit. of H.S.D. @ Rs 4.50	=	Rs 36.00/hour
Lubricants @ 33% of the above	=	Rs 12.00/hour
Sundries @ 10% of the above	=	Rs 3.60/hour
		Rs 51.60/hour

Total Expenses/hour

Depreciation of Tripper	=	56.25
Tyre	=	33.33
Repairs of Tripper	=	78,75
Tyre	=	5.00
W.C. Estt.	=	37.19
P.O.L.s	=	51.60
Total		Rs 262.12/hour

Taking 7 trips of the tripper per hour

$\therefore$ cost/m^3 = $\frac{262.12}{7 \times 4}$ = 9.39 say Rs 9.40/m^3.

3. Working Rate of Bharat D-80A-12 Dozer

Capital cost Rs 20.00 lacs

Life 12,000 hours

Depreciation $\frac{0.90 \times 20,00,000}{12,000}$ = Rs 150/– per hour

Repairs $\frac{2.40 \times 20,00,000}{12,000}$ = Rs 400/– per hour

W.C. Estt. Rs 50/– per hour

P.O.L.'s

18 litres H.S.D. @ Rs 5.00 per lit. Rs 90.00

Sundries @ 43% of the above Rs 38.70

Total Rs 728.70 say Rs 730.00 per hour

4. Working Rate of Batching Plant (25 to 30 m^3/hour capacity)

Capital cost Rs 6,50,000.00

Life 12,000 hours

Depreciation $\frac{0.95 \times 65,000}{12,000}$ = 51.47

Repairs provision $\frac{1.00 \times 65,000}{12,000}$ = 54.17

W.C. Staff

	Rs
1 No. operator @ 1,000/- p.m.	1,000.00
1 No operator for concrete drums @ 1,000/- p.m.	1,000.00
1 No. operator for water pump @ 1,000/- p.m.	1,000.00
3 No. helpers @ 750/- p.m.	2,250.00
1 No. mechanic @ 1,000/- p.m.	1,000.00
½ No. foreman @ 1,200/- p.m.	600.00
½ No. electrician @ 1,000/- p.m.	500.00
1 No. watchman @ 750/- p.m.	750.00
	8,100.00
Add 20% for leave, T.A., bonus, etc.	1,620.00
Total	9,720.00 say Rs 10,000/-

Assuming 1,500 hours of working in one year

Working charges per hour = $\frac{10,000 \times 12}{1,500}$ = Rs 80.00 per hour

P.O.L.'s

7 lit. of diesel @ 4.50 per lit = Rs 31.50

Add for cotton waste. grease etc. @ 25% = Rs 7.90

Rs 39.40

Total Rate

Depreciation	51.47
Repairs	54.17
W.C. Estt	80.00
P.O.L.	39.40
	224.64

Add for expenditure towards erection and dismantling of batching plant Rs 50,000.00 debited to 2 years working only.

$\therefore$ Cost/hour $= \dfrac{50,000}{2 \times 1,500}$ = Rs 16.67 per hour.

Total, including all the above expenses = 324.64 + 16.67 = Rs 341.31 per hour say Rs 340.00

Taking an average output as 20 m^3/hour

$\therefore$ Rate per $m^3 = \dfrac{340.00}{20}$ = Rs 17.00 per m^3

5. Excavator of 1.91 m^3 capacity

Cycle time = 50 seconds

Cycle/hour $= \dfrac{60 \times 60}{50} = 72$

Ideal production = 72 × 1.91 = 138 m^3/hour

Net production = Ideal production × Bucket factor × Efficiency

= 138 × 0.8 × 0.7 = 77.3 m^3

Production/shift = 77.3 × 6 = 464 m^3

Cost = Rs 36 lacs

Life = 12,000 hours

(A) Depreciation $= \dfrac{36,00,00}{12,000} \times 0.95$ = Rs 285/-

(B) Repair provision @ 150% $= \dfrac{1.5 \times 36,00,000}{12,000}$ = Rs 450/-

(C) (i) Diesel 18 litres @ 4.50/litre = Rs 81.00

(ii) 25% lubricants = Rs 20.25

(iii) 3% of (i) & (ii) for storage, etc. = Rs 3.03

(D) Crew (L.S.) = Rs 50.00

Rs 889.28 say Rs 890.00/-

6. Hippo Dumper 8.8 m^3 15 T capacity

1. Cost = Rs 8,00,000.00
2. Life = 10,000 hours
3. Tyre cost = Rs 50,000.00
4. Life of tyres = 2,000 hours

(A) Depreciation $= \dfrac{0.95 \times 8,00,000}{10,000}$ = Rs 76.00

(B) Provision for repair @ 140% $= \dfrac{1.40 \times (8,00,000 - 50,000)}{10,000}$ = Rs 105.00

(C) Tyre replacement

(i) Depreciation, $\frac{50,000}{2,000}$ = Rs 25.00

(ii) Repairs @ 15% = Rs 3.75

Rs 28.75

(D) Provision for work charged staff = Rs 25.00

(E) P.O.L.

(i) Diesel 12 litres @ Rs 4.50/litre = Rs 54.00

(ii) Lubricant, etc., @ 25% = Rs 133.50

(iii) Stock storage @ 3% of (i) + (ii) = Rs 2.05

Rs 69.55

Total = Rs 304.30/hour

7. Dumpers of 18 m^3 capacity

1. Cost of equipment $\simeq$ Rs 9,45,000.00
2. Life = 12,000 hours

Since the work has to be executed in the working seasons only, the depreciation may be limited to an extent of 90% only, and the repair reserve may be provided as 100% as against 150%.

(a) Depreciation $\simeq \text{Rs } \frac{9,45,000}{12,000} \times 0.90 = \text{Rs } 70.87$

(b) Repair Reserve $= \frac{9,45,000}{12,000} \times 1.0 = \text{Rs } 78.75$

(c) Tyre provision $= \frac{15,000}{2,500} = \text{Rs } 6.00$

(d) Repair provision for (c) = 0.40 × 6.0 = Rs 2.40

(e) Crew

(i) One No. operator = Rs 1,000.00

(ii) One No. helper = Rs 750.00

(iii) ¼ Foreman @ Rs 2,000.00 p.m. = Rs 500.00

(iv) ⅙ chowkidar @ Rs 600.00 p.m. = Rs 100.00

Rs 2,350.00

Add 33% for leave reserve, etc. 783.33

Rs 3,133.33 p.m.

Crew charges/hour $= \frac{3,133.33}{6 \times 24}$ = Rs 21.76

(f) P.O.L.

H.S.D. 25 litres @ Rs 4.50/litre = Rs 112.50

Lubricants @ 60% = 75.00

Rs 187.50

∴ Total = 70.87 + 78.75 + 6.00 + 2.40 + 21.76 + 187.50 = 367.28 say Rs 370.00 per hour.

8. Komatsu Bulldozer D-155

Cost = Rs 18 lacs

Life = 12,000 hours

(a) Depreciation @ 90% $= \frac{0.90 \times 18,00,000}{12,000}$ = Rs 135.00

(b) Repair reserve @ 160% $= \frac{1.6 \times 18,00,000}{12,000}$ = Rs 240.00

(c) Cost of tyres $= \frac{15,000}{2,500}$ = Rs 6.00

(d) Repair provision for tyres @ 40% of (c) = Rs 2.40

(e) Operational crew as in the case of item 7 = Rs 21.76

(f) P.O.L.

HSD 30 litres @ Rs 4.50/lit = Rs 135.00

Lubricants 60% = Rs 81.00

Total 135.00 + 240.00 + 6.00 + 2.40 + 21.76 + 135.00 + 81.00

= Rs 621.16 say Rs 620.00 per hour

9. Shovel (3 m^3)

Cost Rs 36.00 lacs

Life 15,000 hours

(a) Depreciation @ 90% $= \frac{36,00,000}{15,000} \times 0.9$ = Rs 216.00

(b) Repair reserve @ 100% $= \frac{36,00,000}{15,000} \times 1$ = Rs 360.00

(c) Crew charges (same as Dozer) = Rs 21.76

(d) P.O.L. (same as in Dozer) = Rs 135.00

Lubricants + 81.00

Total Rs 813.76 per hour say Rs 815.00 per hour

10. Water Tanker

A. Cost of equipment Rs 5.76 lacs

Life 10,000 hours

(i) Depreciation (90% in the total life) $= \frac{5,76,000 \times 0.9}{12,000}$ = Rs 51.74/hour

(ii) Repair reserve @ 150% $= \frac{5,76,000}{12,000} \times 1.5$ = Rs 86.40/hour

(iii) Tyre provision (assuming 2,000 hours as the life of tyres) $\frac{15,000}{2,000}$ = Rs 7.50/hr

(iv) Repair provision of tyres @ 15% = Rs 1.13/hr

Total = 146.77

B. Labour

1/6 foreman @ 18,000.00 p.m. = Rs 300.00

1/6 mechanic @ 1,200.00 p.m. = Rs 200.00

1 operator @ 1,000.00 p.m. = Rs 1,000.00

1⅓ helper @ 750.00 p.m. = Rs 1,000.00

⅙ chowkidar @ 750.00 p.m. = Rs 150.00

Total = Rs 2,650.00

Assuming 1,250 working hours per year

Crew charges per hour $\frac{12 \times 2,650}{1,250}$ = Rs 25.60/hour

C. P.O.L.

18 litres diesel/hour @ 4.50/lit = 81.00

lubricant @ 60% = 48.60

= 129.60

Total = 146.77 + 25.60 + 129.60 = Rs 301.97/hour

say Rs 302.00 per hour

11. Vibratory Compactor

Cost of equipment = Rs 7,56,000.00

Life = 8,000 hours

(a) Depreciation @ 90% = $\frac{7,56,000}{8,000} \times 0.9$ = Rs 85.05

(b) Repair Reserve @ 100% = $\frac{7,56,000}{8,000} \times 1.0$ = Rs 94.50

(c) Labour — same as in the case of item No. 7

Dumpers Rs 21.76/hour

(d) P.O.L.

12 litres/hour @ Rs 4.50/lit = Rs 54.00

Lubricants @ 60% = Rs 32.40

∴ Total = 85.05 + 94.50 + 21.76 + 86.40 = 287.71

Say Rs 288.00/hour

12. Compressor 365 m³/min

Cost of equipment = Rs 2,18,400.00

Life = 8,000 hours

(a) Depreciation @ 90% = $\frac{2,18,400}{8,000} \times 0.9$ = Rs 24.57

(b) Repairs @ 100% = $\frac{2,18,400}{8,000} \times 1.0$ = Rs 27.30

(c) Labour

⅙ foreman @ 1,800.00 p.m. = Rs 300.00

⅓ mechanic @ 1,200.00 p.m. = Rs 400.00

1 operator @ 1,000.00 p.m. = Rs 1,000.00

1 helper @ 750.00 p.m. = Rs 750.00

Rs 2,450.00 p.m.

∴ Crew charges/hour = $\frac{2,450}{6 \times 24}$ = Rs 17.00/hour

(d) P.O.L.

25 lit diesel/hour @ Rs 4.50/lit = Rs 112.50

Lubricants @ 60% = Rs 67.50

Rs 180.00

∴ Total = 24.57 + 27.30 + 17.00 + 180.00 = 248.87 say Rs 249.00/hour

13. Jack Hammer

Cost of equipment = Rs 8,000.00

Life = 3,000 hours

(a) Depreciation @ 90% = $\frac{8,000}{3,000} \times 0.9$ = Rs 2.40

(b) Repair Reserve @ 100% = $\frac{8,000}{3,000} \times 1$ = Rs 2.66

(c) Crew

1/8 mechanic @ Rs 1200.00 p.m. = Rs 150.00

1 operator @ Rs 1000.00 p.m. = Rs 1000.00

Rs 1150.00

∴ Crew charges/hour = $\frac{1150}{6 \times 24}$ = Rs 80.00/hours

Total = 2.40 + 2.66 + 8.00 = 13.06 say Rs 13.10/hour

14. Water Pump 4″ × 4″ (Diesel)

Cost = Rs 35,000.00

Life = 7200 hours

Engine H.P. = 30

(a) Depreciation = $0.95 \times \frac{35,000}{7,200}$ = Rs 4.62

(b) Repairs reserve @ 100% = $\frac{35,000}{7,200} \times 1$ = Rs 4.86

(c) Crew

(i) 1 No. operator @ 1000.00 p.m. = Rs 1000.00

(ii) 1/2 No. helper @ 750.00 p.m. = Rs 375.00

(iii) 1/4 No. chowkidar @ 750.00 p.m. = Rs 187.50

Rs 1,562.50 p.m.

∴ Crew charges/hour = $\frac{1,562.50}{6 \times 24}$ = Rs 10.85/hour

(d) P.O.L.

Diesel 4 litres/hr @ Rs 4.50/lit = Rs 18.00

Lubricants, sundries, etc. @ 25% = Rs 4.50

Stock storage @ 3% = Rs 0.54

Rs 23.04

Total = 4.62 + 4.86 + 10.85 + 23.04 = 43.37 say Rs 43.50/hr

15. Excavator 0.9 m^3, Capacity

Cost = Rs 9,21,265.00

Life = 12,000 hrs

(a) Depreciation @ 90% = $\frac{9,21,265.00}{12,000} \times 0.9$ = Rs 69.09

(b) Repair reserve @ 150% = $\frac{9,21,265.00}{12,000} \times 1.5$ = Rs 115.16

(c) Crew charges

1 No. operator @ Rs 1,000.00 p.m. = Rs 1,000.00

1 No. helper @ Rs 750.00 p.m. = Rs 750.00

¼ No. foreman @ Rs 1,800.00 p.m. = Rs 450.00

⅙ No. chowkidar @ Rs 750.00 p.m. = Rs 150.00

∴ Crew charges/hr $\frac{2350}{6 \times 24}$ = Rs 16.30

(d) P.O.L.

14 litres HSD @ Rs 4.50/lit = Rs 63.00

Lubricants @ 60% = 37.80

100.80

Total 69.09 + 115.16 + 16.30 + 100.80 = Rs 301.35 say Rs 301.50/hour

16. Transit Mixer 6 m^3

Cost Rs 6,75,000.00

Life 10,000 hours

(a) Depreciation @ 90%/hr = $\frac{6,75,000}{10,000} \times 0.9$ = Rs 60.75

(b) Repair Reserve @ 120% = $\frac{6,75,000}{10,000} \times 1.20$ = Rs 81.00

(c) Crew charges as per item No. 15 = Rs 16.30

(d) P.O.L.

Diesel 30 lit/hr @ Rs 4.50/lit = Rs 135.00

Lubricant @ 60% = Rs 216.00

Total = 60.75 + 81.00 + 16.30 + 216.00 = Rs 374.05/hr say Rs 374.00/hour

3

MISCELLANEOUS ITEMS

1.0 Damp-proof Coarsing

1. 2 cm thick damp-proof coarse with cement and approved coarse sand 1 : 2, with and including water-proofing material as ordered by the Engineer I/C in the proportion as specified by the manufacturer, including supply of all materials, labour and T & P, etc., required for proper completion of the work and including proper centering and shuttering, as necessary.

A. Materials — Analysis per 100 sq. ft.

	Unit	Qty.	Rate	Amount
Coarse sand	In cubic metre	0.178 m^3	Rs 200/m^3	35.60
Cement	In 50 kg bags	2.6 bags	Rs 70/bag	162.00
Water proofing compound as specified by the manufacturer	In kg	3.4 kg	Rs 10/kg	34.00
				231.60

B. Labour

	No.	Rate		Amount
Mistri	1/20	Rs 40/day	=	2.00
Mason	1/4	Rs 40/day	=	10.00
Beldar	1/4	Rs 20/day	=	5.00
Coolie	1/4	Rs 20/day	=	5.00
Bhisti	1/8	Rs 24/day	=	3.00
Sundries			=	2.00
			=	27.00
Total 231.00 + 27.00			=	Rs 258.00
Contractor's profit @ 10% on items excluding cement and sand on Rs 51.00			=	5.10
				Rs 263.10 per 100 sq. ft.
			=	Rs 24.50 per sq. metre

2.0 Wood Work

2. C.P. teak wood wrought framed & fixed including simple mouldings, straight works as chaukhats, including supply of all materials, labour and T & P, etc., required for completion of work.

Take a size of opening as 4′ × 7′, section of chaukhat as 4″ × 3″

Wood — 1 × 18 × 1/3 × 1/4	= 105 c.ft.		
Wastage @ 5%	= 0.075 c.ft		
Total	= 1.575 c.ft say 1.6 c.ft.		
Wood	1.6 cft	Rs 120.00/cft	Rs 192.00
Carpenter Mistri	1/16 No.	Rs 40/day	Rs 2.50
Carpenter	1/2 No.	Rs 40/day	Rs 20.00
Sundries L.S.			Rs 10.00
			Rs 224.50
Contractor's profit @ 10%			Rs 22.45
			Rs 246.95

or Rs $\frac{247.00}{1.5}$ = Rs 164.66 per c.ft or Rs 5,800 per cu. m.

3. Supplying and fixing of 4.5 cm (1¾″) thick C.P. teak wood part panelled and part glazed doors and window shutters, including supply and fixing of wooden cleats and stops and including fixing and adjustment of hinges, bolts, locks, handles, springs and other fittings, but excluding their supplies and including painting with 2 coats of black Japan on all iron fittings, if necessary and excluding fixing the glass panes with putty and brad. Glass to be 0.64 kg. Fittings with necessary screws to be supplied departmentally.

Taking a door size 4′ × 7′, size of shutter 3′ 2¾″ × 6′ ¾″ = 24.75 sq. ft.

Wood required including 10% wastage	3.25 cft	Rs 120/cft	Rs 390.00
Glass panes 12″ × 8″	16 No.	Rs 2/each	Rs 32.00
Putty and brad L.S.			Rs 3.00
Mistri	1/8 No.	Rs 40/day	Rs 5.00
Carpenter	4 No.	Rs 40/day	Rs 160.00
Beldar	1 No.	Rs 20/day	Rs 20.00
Glazier	1/4 No.	Rs 30/day	Rs 7.50
Sundries such as glue, black Japan, etc.			Rs 5.00
			Rs 622.50
Contractor's profit @ 10%			Rs 62.25
Total			684.75

Rate per sq. ft. $\frac{684.75}{24.75}$ = Rs 27.70/sq. ft. or Rs 300.00 per sq.m.

4. Additional cost for fixing glass panes with C.P. teak wood beading and felt as directed by engineer in charge, instead of putty including cost of all materials, labour and T & P, etc., required for proper completion of the work.

Taking 10 Nos. glass panes C.P. Teak Wood 5 × 1/3 × 1/32	= 0.05 cft
Wastage 10%	= 0.005 cft.
	= 0.055 cft. say 0.06 cft.

Wood	per cft	0.06 cft	@ 120.00/cft	=	7.20
Screws	per gross	10 doz	@ 12.00	=	10.00
Carpenter	per day	¼ No.	@ 40.00	=	10.00
Beldar	„	⅛ No.	@ 20.00	=	2.50
					Rs 29.70
Contractor's profit @ 10%				=	2.97
					Rs 39.67

16 glass panes measured in item as above = 24.75 sq.ft

So rate per sq.ft = $\frac{16 \times 32.67}{10 \times 24.75}$ = Rs 2.15

∴ Rate per sq. metre = Rs 24.50

5. Supply and fixing of 2.5 cm (1″) thick deodar wood planks, ceiling, including supply of beading 2.5 cm × 4 cm (1″ × 1½″) of deodar wood over joints, but excluding cost of ceiling joist, if and when necessary, but including cost of all materials and labour and T & P, etc., required for proper completion of the work. Cost of 300 sq. ft of ceiling for a room 20′ × 15′.

Planks 1″ thick 300 × 1⁄12	= 25.0 cft
Add for rebates 20%	= 5.0 cft
Beading 400 × 1⁄12 × ⅛	= 4.20 cft
Add for overlap at 10%	= 0.42 cft
Add for wastage 5% on 34.62	= 1.73 cft
Total	= 36.35 cft

Materials	Unit	Qty.	Rate	Cost
Deodar planks	per c.ft	36.35	100.00	3,635.00
Screws 1½″ long	per gross	3 gross	30.00	90.00
Carriage of materials L.S.				30.00
Labour				
Mistry	per day	2 Nos	40.00	80.00
Carpenter	„	2 Nos	40.00	80.00
Beldar	„	4 Nos	20.00	80.00
Scaffolding L.S.				50.00
Sundries L.S.				20.00
				4,065.00
Contractor's profit @ 10%				406.50
				Rs 4,471.50

Cost per sq.ft = $\frac{4,471.50}{300}$ = Rs 14.905

Cost per sq. metre = Rs 160.00

6. Supply and fixing AC plain sheets ceiling including beading 2.5 cm × 4 cm (1″ × 1½″), Deodar wood, screws, bolts and all materials, labour and T & P, etc., complete excluding cost of ceiling joists, if any required, for proper completion of the work.

Take a room 20′ × 15′ = 300 ft^2.
Wastage 5% = 15 ft^2
Total = 315 ft^2
Take 5 ft sq. pane for beading
5 × 20 + 4 × 15 = 160
Wood required 160 × 1/8 × 1/12 = 1.67
Add wastage @ 10% = 0.17
Total = 1.84 say = 2.0 cft
Material

			Rate	Cost in Rs
A.C. sheets	per s.ft	315 s.ft	Rs 3.00/s.ft	945.00
Deodar wood	per cft	2 cft	Rs 125.00/cft	250.00
Screws 1½″	per gross	3 gross	Rs 40.00/gross	120.00
Carriage of materials L.S.				30.00
Labour				
Carpenter	per day	2 Nos	40.00	80.00
Beldar	,,	2 Nos	20.00	40.00
Scaffolding L.S.				25.00
Total				1,490.00
Contractor's profit @ 10%				149.00
				1,639.00
Sundries L.S.				100.00
				1,739.00

∴ Rate per sq. ft = $\frac{1,739}{300}$ = Rs 5.79 say Rs 5.80

or Rs 62.40 per sq. metre

7. 2.5 cm (1″) thick 1 : 2 : 4 plain cement concrete floor with cement, approved coarse sand and 2 cm (¾″) graded approved stone laid in panels flushed with 3 mm (⅛″) flooring coat of neat cement or cement and marble dust in 5 : 1 as specified, and excluding 8 cm (3″) thick base concrete consisting of cement, local sand and brick ballast 4 cm (1½″) gauge in the proportion of 1 : 4 : 8 and removing any overlapping mortar at the joints of panels, if any, and giving them a uniform finish including supply of all materials, labour and T & P, etc., required for proper completion of the work.

Considering 100 ft^2 of the area

Stone grit 2 cm gauge	per % cft	7.5 c.ft	800.00	60.00
Coarse sand	,,	3.7 cft	500.00	18.00
Cement including wastage	per 50 kg	2.4 cft or 2.03 bags	60.00	121.00
Mistri	per day	⅛ No.	40.00	5.00
Mason	,,	1 No.	40.00	40.00
Beldar	,,	½ No.	20.00	10.00
Coolie	,,	½ No.	20.00	5.00
Bhisti	,,	¼ No.	20.00	5.00
				270.30
Adding 1½% water charges				4.05
Sundries and Planties, etc., L.S.				5.00
Contractor's profit @ 10% on items excepting cement, sand and ballast				7.00
				286.35

In this the cost of 4 cm (1½″), 7½ cm thick base concrete is also to be added. The cost can be worked out.

Brick Ballast	% cft	100 cft	350.00	350.00
Fine sand of 1.25 (F.M.)	% cft	50 cft	150.00	75.00
Cement	per 50 kg bag	9.3 bags	60.00	558.00
				983.00
Add for wastage				17.00
				1000.00
Mistri	per day	⅛	40.00	5.00
Mason	"	⅛	40.00	5.00
Beldar	"	3	20.00	60.00
Beldar	"	3	20.00	60.00
Bhisti	"	1	20.00	20.00
Blacksmith	"	⅛	40.00	5.00
Total				155.00
T & P and sundries, etc., hire charges of C.C. mixer; lubricant, etc., including wastage of material @ 2½%				28.87
1½% water charges				17.32
Add contractor's profit @ 10% except on cost of cement, brick ballast and sand				15.50
				1216.69 say 1216.70

Rs 1,216.70 per 100 cft or Rs 369.00 per cu. m.

Therefore the cost of 3″ thick base course would come to Rs 304.17 in 100 sq.ft area.

Total cost per 100 sq.ft = 286.35 + 304.17 = Rs 590.52

or Rs 54.50 per sq.m.

8. First class brick on edge floor laid in cement mortar consisting of one part of cement and six parts of fine sand of 1.25 F.M. and pointed with cement mortar 1 : 2 over and including 8 cm brick ballast base concrete with kankar lime, including supply of all materials, labour and T & P, etc., required for proper completion of work.

Rate per % s.ft — 3″ base concrete in kankar

(a) Base Concrete

Concrete with 4 cm (1½″) gauge, brick ballast and kankar lime in foundations and under floors, including supply of all materials, labour and T & P, etc., required for proper completion of work.

Materials

Brick ballast	% cft	100.00 cft	350.00	350.00
Kankar lime	% cft	33.00 cft	450.00	150.00
				Rs 500.00
Labour				
Mistri	per day	⅛	40.00	5.00
Mason	"	⅓	40.00	13.33
Beldar	"	3	20.00	60.00
Coolie	"	3	20.00	60.00

Bhisti	,,	1	20.00	20.00
Blacksmith	,,	1/8	40.00	5.00
T & P and sundries, L.S.				7.50
				170.83
Materials				500.00
Labour				170.83
				670.83
Add 1½% water charges				10.06
				680.89
Add contractor's profit @ 10% except on cost of brick ballast				33.08
				713.97

Rs 713.97 per 100 cft

or Rs 216.00 per cu. m.

(b) Raised pointing with cement and fine sand of 1.25 F.M. mortar on brick work in proportion of 1 : 2 including supply of all materials, labour and T & P, etc., required for proper execution of work.

Racking and washing surface :				
Mason	per day	1/4 No.	40.00	10.00
Coolie	,,	1/4 No.	20.00	5.00
Bhisti	,,	1/10	20.00	2.00
Labour for pointing				
Mason	,,	1 1/13	40.00	53.00
Coolie	,,	1 1/13	20.00	21.50
Bhisti	,,	1/3	20.00	6.67
Scaffolding L.S.				2.50
Mortar including wastage				
Cement	per 50 kg bag	1 bag	60.00	60.00
Fine sand of 1.25 F. M.	% cft	1 1/14 cft	150.00	160.00
				311.50
Add 1½% water charges on Rs 311.51				4.67
				316.18
Add 10% contractor's profit except on cost of cement and sand				9.54
Total				Rs 325.72

Therefore Rs 325.72 per % sq. ft

or Rs 30.20 per sq. m

(c)

I class bricks	per ‰ No.	530	500.00	265.00
Cement	per 50 kg bag	1.25	60.00	75.00
Fine sand of 1.25 F.M.	per % cft	8.5 cft	150.00	12.75
Labour				
Mistri	per day	1/10 No.	40.00	4.00

Mason	"	1 No.	40.00	40.00
Beldar	"	1 No.	20.00	70.00
Coolie	"	1 No.	20.00	70.00
Bhisti	"	¼ No.	20.00	5.00
				437.75
Add 1½% water charges				6.56
				444.31
Add 10% contractor's profit except on cost of cement and sand				35.65
Total				Rs 479.96

(i) Cost of brick-laying	479.96
(ii) Base concrete	778.44
(iii) Pointing	325.72
	Rs 1584.12

∴ Rs 1584.12 per 100 sq. ft or Rs 147.00 per sq. metre.

9. Granolithic marble flooring 2.5 cm (1") thick, consisting of one part cement two parts of approved coarse sand and 4 parts of 2 cm (¾") graded approved stone chips, including 6 mm (¼") thick surface coat with ordinary cement black, white or grey or mixed marble chips, passing 3 mm (⅛"), but retained on 1.5 mm (1/16") mesh screen in proportion of 1 : 2 as specified or directed by Engineer-in-charge and polished by hand or machine or finally finished with French manson polish or other approved polish, excluding base concrete 8 cm (3") thick. The rates include supply of all materials, labour and T & P, etc., required for proper completion of the work.

¾" C.C. stone chips ¾" granite	per % cft	7 cft	900.00	63.00
Coarse sand	"	4 cft	650.00	26.00
Cement	per 50 kg bag	1.75 bag	60.00	105.00
¼" granolithic chips	per quintal	1.4 quintal	80.00	112.00
Marble chips cement	per 50 kg. bag	1 bag	60.00	60.00
Sundries for rubbing and polishing L.S.				40.00
Mistri	per day	⅛ No.	40.00	5.00
Mason	"	2 No.	40.00	80.00
Beldar	"	2 No.	20.00	40.00
Mistri	"	¼ No.	40.00	10.00
Beldar for cutting and polishing	"	15 No.	20.00	300.00
Sundries L.S. like phanties, etc.				40.00
				881.00
Add 1½% water charges				13.20
				894.20
Add contractor's profit @ 10% (except on the cost of cement, sand and ballast)				52.82
				946.02

∴ Rate per sq. ft = Rs 9.46

or Rate per sq. m = Rs 102.17

Flooring as above, but when white cement is used in top 5 mm (¼") layer instead of ordinary cement.

Rate per % sq. ft as per above				Rs 946.03
Deduct cost of grey cement	per 50 kg bag	1 bag	60.00	60.00
Add cost of white cement	per 50 kg bag	1 bag	150.00	150.00
diff				90.00

∴ Rate 100 sq.ft = Rs 1036.03

∴ Rate per sq. m = $\frac{1036.03 \times 10.8}{100}$ = Rs 112.00 per sq.m.

10. Painting or varnishing on iron works in small areas or new wood work with one coat of ready mixed pointing, paint and one coat of approved superior quality ready mixed paint of clear synthetic varnish including supply of all material, labour and T & P, etc., required for proper completion of work. Paint to be used should conform to the ISI No. 103-1950. Details of cost for an area of 650 s.ft.

Primer	per gal	1 gal	108.00	108.00
Painter	per day	1½ No.	40.00	60.00
Coolie	„	1½ No.	20.00	30.00
Brushes, T & P and other sundries L.S.				10.00
				208.00
Add contractor's profit @ 10%				20.80
				Rs 228.80

Rate per 100 ft^2 = $\frac{228.80 \times 100}{650}$ = Rs 35.00 per 100 ft^2

Details of cost for 650 sq. ft (painting one coat over priming coat).

Paint	per litre	4 litre	50.00	200.00
Painter	per day	1¾ No.	40.00	70.00
Coolies	„	1¾ No.	20.00	35.00
Sundries, brushes, etc., L.S.				10.00
				315.00
Add contractor's profit @ 10%				31.50
				Rs 346.50

Rate per 100 ft^2 = $\frac{346.50 \times 100}{60}$ = Rs 53.20 per 100 ft^2

Hence total cost = 35.00 + 53.20 = Rs 88.20 per 100 ft^2

or Rs 8.20 per sq. metre

11. White washing one coat, including supply of all material labour and T & P required for proper completion of work.

Materials				
White lime	per kg	1 kg	1.00	1.00
Carriage of lime L.S.				0.25
Labour				
White washer	per day	1⁄15	30.00	2.00
Helper (coolie)	„	1⁄15	20.00	1.33
Sundries, gum, ladder, etc., L.S.				1.00

				5.55
Contractor's profit @ 10%				0.55
				Rs 6.10

Rs 6.14 per % sq. ft or Rs 0.58 per sq. m.

12. As in item above but two coats.

White lime	per kg	2 kg	1.00	2.00
Carriage of lime L.S.				0.50
White washer	per day	1/10	30.00	3.00
Helper (coolie)	per day	1/10	20.00	2.00
Sundries, gum, ladder (L.S.)				1.50
				9.00
Contractor's profit @ 10%				0.90
Total				9.90

Rs 9.90 per % sq. ft. or Rs 0.92 pet sq. metre

13. Colour-washing one coat, using approved colouring materials of buff, grey or brown shade, including supply of all materials, labour & T & P etc. required for proper completion of work.

Rate for white-washing one coat				6.14
Add cost of 0.25 kg of colouring materials	per kg	0.25 kg	15.00	3.75
Add extra labour required L.S.				1.00
Contractor's profit @ 10% of Rs 4.75				0.47
				11.36

Rs 11.36 per % sq. ft or Rs 1.04 per sq. metre.

14. As in item above, but 2 coats.

Rate for white washing 2 coats				9.90
Add cost of ½ kg. of colouring materials	per kg	½ kg	15.00	7.50
Add extra labour req. L.S.				2.00
Contractor's profit @ 10% on Rs 9.50				0.95
				20.35

Rs 20.35 per % sq. ft or Rs 1.98 per sq. metre.

15. Cement washing, including supply of all materials, labour and T & P, etc., required for proper completion of work.

Material				
Cement	per 50 kg bag	3.6 kg	70.00	5.04
Carriage of cement L.S.				0.50
Labour				
White washer	per day	1/12 No.	30.00	2.50
Helper (coolie)	"	1/12 No.	20.00	1.66
Sundries, gum, ladder, etc., L.S.				1.00
				10.70
Contractor's profit @ 10%				1.07
Total				11.77

Rs 11.77 per % sq. ft or 1.08 per sq. metre.

16. One priming coat and one coat of dry distemper on new work, including supply of all material, labour and T & P, etc., required for proper completion of the work.

Details of cost of 100 sq. ft.

Whiting	per kg	1 kg	2.00	2.00
Vinegar L.S.				1.50
Labour				
Painter	per day	⅛ No.	40.00	5.00
Coolie	"	⅛ No.	20.00	2.50
Distempering				
Dry distemper	per kg	¾ kg	36.00	27.00
Painter	per day	⅕ No.	40.00	8.00
Coolie	"	⅕ No.	20.00	4.00
Sundries L.S.				4.50
				54.00
Contractor's profit @ 10%				5.40
Total				59.40

Rs 59.00 per % sq. ft or Rs 5.50 per sq. metre.

17. As in item above, but for each coat of distemper after the first coat.

Distemper	per kg	0.6 kg	36.00	21.00
Painter	per day	⅙ No.	40.00	6.66
Coolie	"	⅙ No.	20.00	3.33
Sundries L.S.				3.00
				34.00
Contractor's profit @ 10%				3.40
				38.04

Rs 38.04 per % sq. ft or 3.54 per sq. metre.

18. Dismantling sun-dried brick work including stacking of dismantled materials as directed by Engineer-in-charge within a distance of 60 metres (200 ft).

Coolie	per day	¾ No.	20.00	15.00
Contractor's profit @ 10%				1.50
				16.50

Rs 16.50 per % cft or Rs 5.75 per cubic metre.

19. Dismantling brickwork or stonework in lime or cement mortar and including stacking of materials as directed by the Engineer- in-charge within a distance of 60 m (200 ft).

Beldar	per day	2 Nos	20.00	40.00
Coolie	"	2 Nos	20.00	40.00
				80.00
Contractor's profit @ 10%				8.00
				88.00

Rs 88.00 per % cft or Rs 25.00 per cubic metre.

20. As in above, but in mud mortar.

Coolie	per day	1 No.	20.00	20.00
Contractor's profit @ 10%				2.00
				22.00

Rs 22.00 per % cft or Rs 7.70 per cubic metre.

21. Dismantling reinforced cement concrete or reinforced brick including stacking of dismantled materials as directed by the Engineer-in-charge within a distance of 60 m (200 ft).

Beldar	per day	4 Nos	20.00	80.00
Coolie	,,	4 Nos	20.00	80.00
Blacksmith	,,	⅓ No.	30.00	10.00
Sundries L.S.				15.00
				185.00
Contractor's profit @ 10%				18.50

Rs 203.50 per % cft or 70.10 per cubic metre.

22. Dismantling corrugated iron roof walls and partitions including purlin battens and vertical supports and including stacking of materials as directed by the Engineer-in-charge within a distance of 60 m (200 ft).

Blacksmith	per day	½ No.	30.00	15.00
Coolie	,,	½ No.	20.00	10.00
				25.00
Contractor's profit @ 10%				2.50
				27.50

Rs 27.50 per % cft or Rs 9.45 per cubic metre.

23. Dismantling of girders, trusses and other steelwork including stacking of dismantled materials as directed by Engineer-in-charge within a distance of 60 m (200 ft).

Blacksmith	per day	½ No.	30.00	15.00
Beldar	,,	1½ No.	20.00	30.00
Sundries L.S.				10.00
				55.00
Contractor's profit @ 10%				5.50
				60.50

Rs 60.50 per % cft or Rs 21.20 per cubic metre.

24. Dismantling doors and windows, including chowkhats, with holdfasts and including stacking of dismantled materials, as directed by the Engineer-in-charge within a distance of 60 m (200 ft).

Mason	per day	1/10 No.	40.00	4.00
Carpenter	,,	½ No.	40.00	20.00
Coolie	,,	1 No.	20.00	20.00
				64.00
Contractor's profit @ 10%				6.40
				70.40

Rs 70.40 per % sq. ft or Rs 7.60 per sq. metre.

25. Dismantling doors and window leaves including stacking of materials as directed by the Engineer-in-charge within a distance of 60 m (200 ft).

Carpenter	per day	⅓ No.	40.00	13.33
Beldar	,,	⅓ No.	20.00	6.67
Sundries L.S.				4.00
				24.00
Contractor's profit @ 10%				2.40
				26.40

Rs 26.40 per % sq. ft or Rs 2.85 per sq. metre.

26. Dismantling doors and windows including chowkhats but excluding holdfasts and including stacking of materials as directed by the Engineer-in-charge within distance of 60 m (200 ft).

Carpenter	per day	⅓ No.	40.00	13.33
Mason	,,	18 No.	40.00	5.00
Coolie	,,	1 No.	20.00	20.00
Sundries L.S.				4.00
				42.35
Contractor's profit @ 10%				4.20
				46.55

Rs 46.56 per % sq. ft or Rs 5.05 per sq. metre.

27. Dismantling existing plaster including disposed of refuse as directed by the Engineer-in-charge within a distance of 60 m (200 ft).

Coolie	per day	1 No.	20.00	20.00
Bhisti	,,	1/16 No.	40.00	2.50
				22.50
Contractor's profit @ 10%				2.25
				24.75

Rs 24.75 per % sq. ft or Rs 2.67 per sq. metre.

28. Racking out joints of brickwork in cement or lime mortar to a depth of 1 cm (⅜") including disposal of refuse as directed by the Engineer-in-charge within a distance of 60 m (200 ft).

Beldar	per day	¾ No.	20.00	15.00
Contractor's profit @ 10%				1.50
				16.50

Rs 16.50 per % sq. ft or Rs 1.78 per sq. metre.

29. Scrapping old white or colour wash including disposal of refuse as directed by the Engineer-in-charge within a distance of 60 m (200 ft).

Coolie	per day	⅓ No.	20.00	6.67
Contractor's profit @ 10%				0.66
				7.33

Rs 7.33 per % sq. ft or Rs 0.80 per sq. metre.

30. Consolidation of kankar by hand rammers including open out stacks, sorting Kankar into three sizes, repairing patch in existing road surface with kankar, picking up old surface, spreading kankar, making medhi and ramming and also including carriage of metal upto 30 m (100 ft). Carriage of water 200 m, supply and maintenance of barriers, caution posts, light at night and including all T & P, etc., required for proper completion of work. Kankar to be supplied to the contractor free at site.

Taking up a length of 220′ × 12′ = 2640 sq.ft

Kankar required 2640 × 3⁄8 = 990 cft.

Beldar for crisscross lines and sweeping and take out kankar dug	per day	2 Nos	20.00	40.00	
Preparing earth and making medhi mazdoor	,,	1 No.	20.00		20.00
Beldar for opening out stacks and sorting into sizes	,,	2 Nos	20.00		40.00
Mazdoor	,,	2 Nos	20.00		40.00
Mazdoor for carrying of kankar and spreading on surface	,,	4 Nos	20.00		80.00
Hand packing beldar	,,	6 Nos	20.00		120.00
Supply of water, mazdoor	,,	8 Nos	20.00		160.00
Consolidation, head rammer man	,,	1 No	30.00		30.00
Rammer man	,,	11 No.	20.00		220.00
Leak karai L.S.					15.00
T & P night light L.S.					3.00
					708.00
Contractor's profit @ 10%					70.00

Rate for % cft = $\frac{888.80}{990} \times 100$ = Rs 89.50

or Rs 31.40 per cubic metre.

31. As in item above but for consolidation by Road Roller including hire and fuel of the roller.

As in item above				888.80
Deduct for labour involved in consolidation				290.00
Add for hire charges of roller	per day	1 No.	80.00	80.00
Steam coal	per ton	1⁄4 T	500.00	125.00
Beldar	per day	2 No.	20.00	40.00
Contractor's profit except on roller and coal				4.10
				847.50

Rate per % cft = $\frac{847.50}{990} \times 100$ = Rs 86.50

or Rs 9.30 per cubic metre.

32. Bricksoling with straight overburnt bricks laid flat and dry, including preparations of sub-grade to prescribed camber and including supply of bricks, filling joints with local earth, spreading a blinding coat of 1 cm (3⁄8″) depth of local earth of P.I. 5 to 7 or sand over the entire surface and including labour and T & P, etc., required for proper completion of the work.

Bricks	‰ Nos	350 Nos	500.00	175.00
Rate of preparing sub-grade (100 sq. ft)				
Mazdoor	per day	1 No.	20.00	20.00
Laying bricks — mason	,,	1/6 No.	40.00	6.66
Mazdoor	,,	1/2 No.	20.00	10.00
Spreading earth — Mazdoor	,,	1/2 No.	20.00	10.00
				221.66
Contractor's profit @ 10%				22.16
				243.82

Rs 243.82 per 100 sq. ft or Rs 26.30 per sq. metre.

33. As in item above, but with bricks laid on edge.

Bricks	‰ No.	550 No.	500.00	275.00
Rate of preparation of sub-grade				
Mazdoor	per day	1 No.	20.00	20.00
Laying bricks — mason	,,	11/30 No.	40.00	14.67
Mazdoor	,,	1/2 No.	20.00	10.00
Spreading earth — mazdoor	,,	1/2 No.	20.00	10.00
				429.67
Contractor's profit @ 10%				42.96

Rs 472.63 per 100 ft^2 or Rs 51.00 per sq. metre.

34. Bitumen painting (labour only) over W.B. surface (I coat) excluding supply of paint and grit including maintenance of side patries and watering them during operation for proper completion of the work as directed by the Engineer-in-charge and excluding roller charges and its fuel but including the wages of fireman and tar boiler man.

Take a strip of 1/2 furlong area 330 × 12 = 3960 sq. ft.

Cleaning gait mazdoor	per day	1 No.	20.00	20.00
Cleaning sub-grade mazdoor	,,	10 Nos	20.00	200.00
Coolie	,,	8 Nos	20.00	160.00
Tar heating and spreading on road				
Hot beldar	,,	7 Nos	30.00	210.00
Grit spreading beldar	,,	5 Nos	25.00	125.00
Diversion Watering				
Beldar	,,	2 Nos	25.00	50.00
Misc. Mate	,,	1 No.	30.00	30.00
Work agent	,,	1 No.	50.00	50.00
Tar boiler and fireman	,,	1 No.	40.00	40.00
Sundries L.S.				100.00
				985.00

$$\text{Rate \% sq. ft} = \frac{985.00}{39.60} = 24.80$$

Rate per sq. metre = Rs 2.68.

35. Laying C.C. road Slab 10 cm (4″) thick, consisting of 1 : 2 : 4 cement, coarse sand and approved 4 cm gauge stone ballast, over prepared sub-grade after its rectifications and bringing it to proper camber and including supply of all materials, labour and T & P, etc., required for proper completion of work, finish to required template as directed by Engineer-in-charge, but excluding cost of metal required for rectification of sub-grade.

Take a slab length of 33 r.ft and 4″ in depth. Quantity of work involved = 33 × 12 × ⅓ = 132 cft.

Stone ballast 1½″ gauge	per % cft	119 cft	1500.00	1790.00
Coarse sand	,,	60 cft	1300.00	780.00
Cement bags	per 50 kg bag	25 Nos	70.00	1750.00
Empty bags	each	15 Nos	1.00	15.00
Tar and other side forma, etc., L.S.				80.00
Laying of C.C. and putting forma and temping, etc.				
Mistri	per day	⅙ No.	48.00	8.00
Mason	,,	2 Nos	20.00	40.00
Mate	,,	1 No.	30.00	30.00
Beldar	,,	2 No.	20.00	40.00
Coolie	,,	2 Nos	15.00	30.00
Sub-grade rectification, etc.				
Mason	,,	½ No.	40.00	20.00
Beldar	,,	1 No.	20.00	20.00
Curing				
Mate	,,	1 No.	30.00	30.00
Coolie	,,	8 Nos	15.00	120.00
Diversion				
Beldar	per day	1 No.	20.00	20.00
Chowkidar	,,	2 Nos	20.00	40.00
Misc. supervision				
Work agent	,,	⅓ No.	50.00	16.67
Driver C.C. Mixer, etc.	,,	⅙ No.	40.00	6.67
Cleaner	,,	⅙ Nos	20.00	3.33
Beldar	,,	3 Nos	20.00	60.00
Coolie	,,	3 Nos	15.00	45.00
				4944.67
Water Charges @ 1½%				72.00
Contractor's profit @ 10% on Rs 763.00 (labour charges only)				76.30
				5092.97

$$\text{Rate for 1 cft} = \frac{5{,}092.97}{132} = \text{Rs } 38.58 \text{ (app.)}$$

or Rs 705.00 per cubic metre.

36. Earthwork in edging of the painted or C.C. miles in a width of ½ m (20″) on either side, according to Patri template, including dressing ramming, etc. The trenches to be so dug for taking the earth that they form a continuous side drain at road boundary.

Take a length of one furlong. Earthwork to be done (average) say 380 cft.

Rate for earth work	per ‰ cft	380 cft	100.00	38.00
Add 50% for small work				19.00
				57.00

Rs 57.00 per furlong or Rs 456.00 per mile or Rs 285.00 per km.

37. Joint and small cracks filling in C.C. miles after racking out the joints and removing the old tar and other materials with suitable materials as directed by Engineer-in-charge (labour only).

Total joints at every 33′ joints to be filled would be = $\frac{5280}{33}$ = 160 Nos

Add 25% for other cracks = 40 Nos

Total 200 Nos

Thus 200 joints will be attended to.

Mate	per day	¾ No.	30.00	22.50
Beldar	,,	6 Nos	20.00	120.00
Coolie	,,	1 No.	15.00	15.00
				157.50
Contractor's profit @ 10%				15.75
				173.25

Rs 173.25 per mile or Rs 108.00 per km.

38. Patch repair of painted miles with tar and bitumen (labour only) including heating of paint upto required temperature, pouring it on the surface, spreading the necessary quantity of grit and ramming hard with hand or rammer and also including cleaning with brush, etc., the existing surface after scarifying it to the W.B. surface in rectangular shape and spreading the dismantled material on the patries as directed by Engineer-in-charge.

As per PWD circular, patch work is carried out upto a stage when 1½ tonne tar is used per year per mile. Surface to be treated by 1½ ton roller in a year $\frac{3360}{12}$ = 12,00 s.ft. Grit to be consumed 120 × 2.25 = 270 cft.

Hence consumption per mile per month = $\frac{270}{12}$ = 22.5 say 25.00 c.ft.

Mate	per day	1 No.	30.00	30.00
Beldar	,,	4 Nos	20.00	80.00
Coolie	,,	4 Nos	15.00	60.00
				170.00

Rate for % cft = 680.00 per % cft of grit used.
or Rs 238.00 per cubic metre.

39. Cleaning of existing painted surface and laying 2 cm (¾″) thick over a tack coat with graded approved grit and bitumen (labour only) including heating the paint to the required temperature, mixing either in mixer or in hand drum including carriage of material, maintenance of side patri and watering them during operation for proper completion of work as directed by the Engineer-in-charge. Take a length of ½ furlong 33 × 12 = 3,960 sq. ft.

Labour for cleaning				
Mazdoor	per day	6 Nos	15.00	90.00
Coolie	"	6 Nos	15.00	90.00
Tack coat heating and spreading				
Beldar	"	4 Nos	20.00	80.00
Premixing, screening grit				
Beldar	"	1 No.	20.00	20.00
Heating the tar and mixing grit				
Beldar	"	5 Nos	20.00	100.00
Coolie	"	3 Nos	15.00	45.00
Spreading the premix and making into camber				
Beldar	"	15 Nos	20.00	300.00
Diversion and watering				
Mazdoor	"	2 Nos	15.00	30.00
Miscellaneous Mate	"	2 Nos	30.00	60.00
Work agent	"	1 No.	50.00	50.00
Tar, bitumen and fireman	"	2 Nos	20.00	40.00
Sundries, cartage and other T & P, etc.				
L.S.				100.00
				1005.00

Rate per % ft = $\frac{1,005.00}{39.60}$ = Rs 26.50

or Rs 2.86 per sq. metre.

40. Cleaning the premix surface and laying the seal coat with approved 6 mm (¼″) graded grit and bitumen (labour only) including heating the paint to required temperature, including cartage of material and including maintenance of side patries and watering them during operation as directed by Engineer-in-charge. Take a length of ½ furlong. Area covered 330 × 12 = 3,960 sq. ft.

Labour for cleaning				
Mazdoor	per day	2 Nos	15.00	30.00
Coolie	"	2 Nos	15.00	30.00
Heating and spreading				
Beldar	"	4 Nos	20.00	80.00
Screening graded grit ¼″				
Beldar	"	1 No.	20.00	20.00
Miscellaneous				
Mate	"	1 No.	30.00	30.00
Work agent	"	¼ No.	50.00	12.50
Tar boilerman and fireman	"	1 No.	20.00	20.00
Sundries, cartage and T & P, etc.				25.00
L.S.				247.50

Rate per % sq. ft = $\frac{247.50}{39.60}$ = Rs 6.25 or Rs 0.67 per sq. metre.

41. Cleaning of existing water bound surface and laying premix 2 cm (¾″) thick over a tack coat with graded approved grit & bitumen (labour only) including heating the paint to the required temperature, mixing either in a mixer or in hand drum including cartage of material, maintenance of side patries and watering them during operation for proper completion of work as directed by Engineer-in-charge.

Take a length of ½ furlong area = 330′ × 12′ = 3960 ft²

Labour for cleaning				
Mazdoor	per day	10 Nos	15.00	150.00
Coolie	„	10 Nos	15.00	150.00
Tack coat, heating and spreading				
Beldar	„	6 Nos	20.00	120.00
Premixing, screening grit				
Beldar	„	1 No.	20.00	20.00
Heating and tar mixing grit				
Beldar	„	5 Nos	20.00	100.00
Coolie	„	3 Nos	15.00	45.00
Spreading the premix and making into camber				
Beldar	„	15 Nos	20.00	300.00
Diversion and watering				
Beldar	„	2 Nos	20.00	40.00
Miscellaneous				
Mate	„	3 Nos	30.00	90.00
Work agent	„	1 No.	50.00	50.00
Tar boilerman and fireman, etc.	„	2 Nos	20.00	40.00
Sundries, cartage, T & P, etc.				100.00
L.S.				1,205.00

$$\text{Rate per \% sq. ft} = \frac{1{,}205.00}{39.60} = \text{Rs } 30.50$$

or Rs 3.30 per sq. metre.

42. Maintenance of terraced roofs filling up cracks before rains, with approved bituminous filler and stopping leaks during rains including supply of all materials, labour and T and P, etc. required for proper completion of the work.

Take an area of % sq. ft.

Cleaning coolie	per day	1 No.	15.00	15.00
Bitumen filling and other minor repairs				
Contractor's profit @ 10%				23.50
				38.50

Rs 38.50 per % sq. ft or Rs 41.50 per % sq. metre.

43. Repair to plaster or pointing in patches, including removal of old plaster, racking out joints to 1 cm, (⅜″) depth, watering the area and then replastering or repointing with the same mortar in proportion of mortar as the existing surface including supply of all materials, labour and T & P, etc., required for proper completion of work.

This rate of 25% higher than the normal rate for plaster or pointing as this work is done in patches.

44. Repair to 2.5 cm (1″) cement flooring patches, including digging upto 10 cm (4″) depth renewing 8 cm (3″) base concrete consisting of 1 part cement, 4 parts approved fine sand of 1.25 F.M. and 8 parts of 1½″ gauge brick ballast as in C.C. consisting of one part cement 2 part approved coarse sand and four parts approved stone grit, including supply of all materials, labour and T and P etc. required for proper completion of work.

Add 25% more to the rate for the main item.

45. Fixing panes of glass with and including putty and brads and including supply of all materials, labour and T & P, etc., required for proper completion of work.

Glazer can fix 30 glasses of 1′ × 1 . Hence for 1 sq. ft 1/30 glazer is required.

Glazer	per day	1/30 No.	60.00	2.00
Sundries, putty and brad, etc., L.S.				1.00
Contractor's profit @ 10% on Rs 3.50				0.35
Total				3.85

Rs 3.85 per sq. ft. = Rs 41.50 per sq. metre.

46. Labour for fixing door bolts or hinges or tower bolts or hasp and stoppers of handles etc. to doors and windows. Carpenter can fix 24 such type of doors or windows fitting in a day.

Hence for 1 fitting 1/24 carpenter is required.

Carpenter	per day	1/24 No.	50.00	2.08
Contractor's profit @ 10%				0.20
				Rs 2.28 or Rs 2.30 each.

47. Opening and refixing 10 cm (4″) rain water down, pipes after cleaning including supply of all materials, labour, and T & P, etc., required for proper completion of the work.

Take a length of 6 r.ft.

Carpenter	per day	1/8 No.	50.00	6.25
Coolie	,,	1/8 No.	15.00	1.87
Sundries, nails, etc., L.S.				3.00
				11.12
Contractor's profit @ 10%				1.11
				12.23

Rate per r. ft = Rs $\frac{12.26}{6}$ = Rs 2.04 or Rs 6.70 per r. metre.

4

ESTIMATING

An estimate is an assessment of the quantities of different materials required and the expenditure likely to be incurred in the execution of civil engineering work. Thus the estimate gives probable cost of a work and is determined by mathematical computation of the quantities involved, on the basis of drawings and plans, and finally the cost worked out therefrom on the basis of analysis of rates of different items involved. The actual cost of the work is known only after completion of the work. Obviously the actual cost of the work should not deviate much from the estimated cost.

Estimates can be divided into two categories — approximate and detailed. An approximate estimate is used in making preliminary studies of various aspects of a project.

The approximate estimate of a building can be prepared on the basis of

(a) Service-unit method

(b) Square-metre method

(c) Cubic metre method

(d) Typical bay method

(e) Approximate quantities method

Service Unit Method : The service unit indicates the most important unit in a structure, e.g., in a school building, the service unit is a class room, in a hospital it is a bed. In a water tank it is the unit cubical content expressed in terms of litre, in a cinema house, it is seat. So approximate cost of a structure = Number of service units in the structure × cost of corresponding service units in a similar existing structure. For example, suppose a school building of 16 class rooms is to be constructed, and the cost of construction of a class room in a similar existing school building is Rs 10,000/- Then the approximate cost of the proposed school building would be 16 × 10,000 = Rs 1,60,000.

The variations in the approximate cost worked out on the basis of the above could be considerable due to type of construction and standard of finish; structural dis-similarity; variation in the rates of material and labour at the time of execution.

Square Metre Method : In this method the area of each floor is worked out and added together, to yield the total area of the building. Then the approximate cost of the building = Total area of all the floors × cost per square metre of a similar existing building. For computing floor area, measurements are taken out to out of walls and generally no deductions are made for stairways or similar other gaps, and the projections at the plinth level are excluded. This method will lead to variations in the approximate cost due to change in price level; differing ceiling height; type of construction; shape of building; and the different location of the buildings.

Cubic Metre Method : In this method cost per cubic metre is used as a base for calculating the approximate cost of the proposed building. Cubical contents of the proposed building are worked out and then, approximate cost of proposed building = Total cubical contents of proposed building × Cost per cubic metre of a similar existing structure.

However, there is no general agreement regarding the inclusion of cubical contents of open verandahs, open porches, plazas etc. There is no agreement regarding the height of the building too.

Typical-bay Method : This method may be adopted in case of buildings, which have several similar bays. A bay being the space from centre to centre of two successive columns. A typical interior bay is selected and its total cost is worked out.

Approximate cost of proposed building = Number of bays × Cost of one bay.

Approximate Quantities Method : In this method, various quantities are worked out, such as excavation, foundation concrete, brick work upto plinth level and damp proof course etc. and with the known rates of these items, the cost of building could be worked out. Similarly the quantities in the super structure are worked out, which include brick work, concrete, wood work and finishing items etc. The rates of these items being known before hand, the cost of the building could be worked out.

Detailed Estimation

Detailed estimates are prepared in two stages. In the first instance quantities of different items of work are worked out, and then the rates of these items are evolved on the basis of analysis of rates. Finally the cost of each item is obtained by multiplying the quantity of the item with the respective rate of the item. The cost of each item is added to obtain cost of the work.

Working Out Quantities : The quantities of the various items are worked out on a dimension form and then the quantities of identical items are grouped together and written on an abstract form. A dimension form could conveniently be as below. The columns are self explanatory.

Item No.	Particulars of Item	No.	Length	Breadth	Height	Contents or quantity

The standard modes of measurement for some of the most common items are given below. The readers are advised to refer to I.S. Code 1200-1958.

(i) *Preliminaries :* These are the items required to be done for clearance of the site, such as jungle clearance, removing trees, and demolishing the existing old structures, etc. These jobs are included in the estimates on lump sum basis, unless the quantities of each work are too appreciable.

(ii) *Earthwork :* Earthwork in excavation and filling are calculated separately. The unit for all earthwork is taken as cubic metre. The length and width are measured according to the drawings, with no allowance given for the working space. The depth is measured vertical at a number of points, and a mean value is taken, if there is no wide variation, otherwise the area is divided in suitable sections and different depths are obtained for each section. The measurements of excavation for every 1.5 m depth are taken separately. Similarly for different leads also, measurements are taken separately. If excavated earth is used in branch filling and plinth filling, then it is not paid for separately, but may also be included as a separate item "Returnfill and ram or backfill", and paid at a lower rate.

Concrete : The foundation concrete is measured in cubic metres. The length and width being the same as in excavation, and the depth is measured as per actual concrete laid.

RCC slabs upto 15 cm thick, RCC partition wall, and RCC box are measured in sq. metres. RCC lintels, beams and columns are measured in cubic metres. RCC stair is measured in terms of number of steps. Halfsteps and quarter landing are considered as equivalent to four steps and two steps respectively.

RCC weather shed is measured in sq. metres and the depth of the course is specified.

Precast cement concrete jali work is measured in sq. metres and the thickness is specified. Precast cement concrete, posts, struts, etc., are measured in cubic metres. Centering and shuttering (form work) are usually included in the RCC or RB work, but may also be taken separately in sq. m of surface in contact with concrete.

Masonry : The brick work is measured in cubic metres for one brick wall to three brick walls. For brick of size 19 cm × 9 cm × 9 cm, the thickness of walls for one to three brickwalls shall be respectively 20 cm × 30 cm × 40 cm, 50 cm and 60 cm. If the thickness of the wall is more than three bricks, the actual thickness is measured to the nearest one cm.

The half brick partition wall and the walls having thickness less than half brick are measured in sq. metres.

The brick work for each is measured separately in cubic metres.

It has to be noted that while taking measurements of the brick work, walls are treated as solid walls without any openings. But after the measurements are completed, deductions are made for doors, windows, ventilators, lintels, bearings of beams etc.

Foundation and plinth masonry is taken as one item while the masonry in super structure is taken separately. Masonry of different types or classes, masonry with different mortar, etc. are taken out under separate items. Honey comb brick wall is taken under a separate item in sq. m, no deduction is made for hold fast. Stone masonry is calculated in the same manner as for brick masonry.

Plastering and Pointing : The thickness of the plaster and the mix ratio of the mortar has to be specified. The measurements for plastering are recorded in sq. metres.

(i) If the quality of the plaster is the same on both side of the wall, openings in the wall are deducted only once and no separate account of the sides, bottom and top of the openings is required to be taken.

(ii) If the two sides of the wall are provided with different types of plaster, one half of the area of the openings is deducted from each type of plaster.

(iii) The opening which are more than 3 m^2 in area are to be deducted for two sides separately, and the measurements for the top, bottom and sides of such opening are to be accounted for.

Plastering of ceiling usually of 12 mm thick is computed under a separate head, as this work is done with richer mortar. For RCC work usually no plaster is allowed but for fair finish a thin plaster of rich cement mortar may be allowed, which should not be taken in the measurements separately.

The type of pointing and quality of the mortar are mentioned in the specifications. The measurements are taken of the whole flat surface in square metres. Deductions are made as per items of plastering.

Wood Work : The type of the material used and the quality of finish required should be clearly mentioned in the specifications. The rate for any type of wood work includes fabrication, fixation fastenings, fixtures and three coats of oil paint. The measurements are taken of the net work done and no consideration is given to the wastage of material. However, tolerance of 2 mm is allowed. Rectangular wooden beams and vertical wooden posts are measured in cubic metres. Frames for doors, windows, ventilators and cupboards are measured in cubic metres. Wooden trusses are measured in cubic metres. Wooden door, windows, ventilators and cupboard shutters are measured in sq. metres. Wooden cornices, picture rails, architraves, shelves etc. are measured in running metres.

The name of the timber used, the thickness of shutters, type of shutters and the nature of fittings should be noted in the item. Shutters of different type as panelled, glazed, partly panelled and partly glazed should be computed separately as the rates differ.

Fittings are computed by number, i.e., enumerative fittings may also be included in the sq. m. rate of shutters. These may be taken as separate item for estimation purposes. Similarly holdfasts are taken separate item by weight or by number. In estimates the cost of fittings will be provided under a separate item "fittings of doors and windows" on lump sum basis.

Steel Work : Grill work in windows is measured in sq. metres. Rolling steel shutters steel doors and windows are also paid as per sq. metre. Plain rolled sections, built up works, steel bars, etc. are paid by weight in quintals. For calculating weight of steel members, the density could be taken as 7850 kg/m^3 (490 lbs/cft). Certain percentage of the whole steel work is provided for rivets and bolts and nuts. For steel roof truss 5% of the steel work is usually provided for rivets, bolts and nuts.

White-washing, Colour-washing or Distempering : The surface is calculated in sq.m. Number of coats of white washing or colour washing are taken as one job or work and the rates cover for the number of coats, which should not be a multiplying factor. Number of coats should be mentioned in the item. Corrugated surfaces are measured flat and then a suitable coefficient is applied to compensate for the corrugations. Deductions are made as per item of plastering.

Painting : The item of painting is paid in sq. metres. The measurements are taken flat and then multiplied by a suitable coefficient to compensate for mouldings, rebates, etc. Painting for works less than 15 cm wide, is paid in running metres. No separate measurement is taken for the chowkhat, the area is same as the area of wall opening. For iron bars, grills etc., the area of the clear opening inside the chowkhat is taken. For panelled, framed and braced, ledged and battered or ledged battened and braced, it is customary to take 2½ times one surface area, for both sides. In case of fully glazed or gauged doors and windows only one time one surface area, for both sides is taken.

Painting is done in two or three coats, usually over a coat of priming. The rates cover for the number of coats under one item. The number of coats should be mentioned in the description of item.

Roof Covering : The roof coverings are measured flat without any laps and are paid per sq. metre, and no extra amount is paid for any wastage, cutting etc. Ridge, hips and valleys are measured in running metres.

Flooring : Measurement of floor finishes are taken in sq. metres for the net area covered. No extra amount is paid for the wastage of materials.

Water Supply and Drainage Connections : The item of galvanised iron or cast iron pipe line is measured in running metres for the length laid and no extra amount is paid for the specials, such as bends, tees etc. The measurement is taken along the centre lime of the pipe line. Similarly for drainage the item of pipe line is measured in running metres for the actual length laid and no extra amount is paid for the specials such as bends, tees etc. Manholes, and inspection chambers are paid per number and the specifications should give their full description. Gully traps, nahnitraps, W.C. pans, flushing tanks, mica flap valves, etc. are paid per number and the specification should mention their quality, make, method of fixing etc. For sanitary and water supply works 10% and for electrification another 10% of the cost of the building is provided in the estimate.

In order to achieve the desired degree of accuracy in the estimating work, the dimensions should be measured to the nearest 1 cm, (.01 m), areas should be worked out to the nearest 0.01 sq. m. and cubic contents should be worked out to the nearest 0.01 cu.m. Thickness of slabs, partitions, etc., and sectional dimensions of columns, pillars beams, etc., should be taken to the nearest half cm. (.005 m).

5

METHODS OF BUILDING ESTIMATE

Separate or Individual Wall Method : In this method the external length of walls running in one direction, the long wall out-to-out, and the internal length of transverse walls in-to-in, is taken. The quantities are obtained by multiplying length, width and height of these. The same rule applies to excavation and concrete in foundation. It is convenient to imagine plans at different level of height as foundation trends plan, foundation concrete plan, plans of each footing etc. and dealing each place or part separately.

Centre Line Method : In this method, sum-total length of centre line of all walls, long and short has to be found out. Thus in this method, the length will remain same for excavation in foundation, for concrete in foundation, for all footings and for super structure. This method is quick, but requires special attention and consideration at the junction meeting points of partition or cross-walls.

Illustrative Examples : The method of estimating is illustrated with the help of the following examples.

Example 5.1 Estimate of R.C.C. beam shown in Fig. 5.1.

Item No.	Description of Item	No.	L in m	B in m	H in m	Qty	Den.
1.	**R.C.C. Work 1 : 2 : 4**						
	Excluding steel but including centering and bending of steel. (No deduction for volume of steel bars)	1	8.60	0.40	0.75	2.58	cum
2.	**Steel bar (mild steel)**						
	20 mm ϕ straight bar	5	8.9	–	–	44.5	metre
	$L = 8.60 - \frac{2 \times 5}{100} + \frac{18 \times 22}{1000}$ = 8.896 m $\simeq$ 8.9 m.						
	20 mm bent up bars	4	9.5	–	–	38	metre
	$L = 8.6 - \frac{2 \times 5}{100} + \frac{18 \times 20}{1000} + \frac{.84 \times 60}{1000}$ = 9.406 $\simeq$ 9.5 metre						
	Total length of 22 mm ϕ bars = 82.5 m						
	Wt. = $7.5 \times \frac{\pi}{4} \times (2.2)^2 \times \frac{1}{100 \times 100} \times 82.5^{10}$ = 2.35 quintal						

Continued

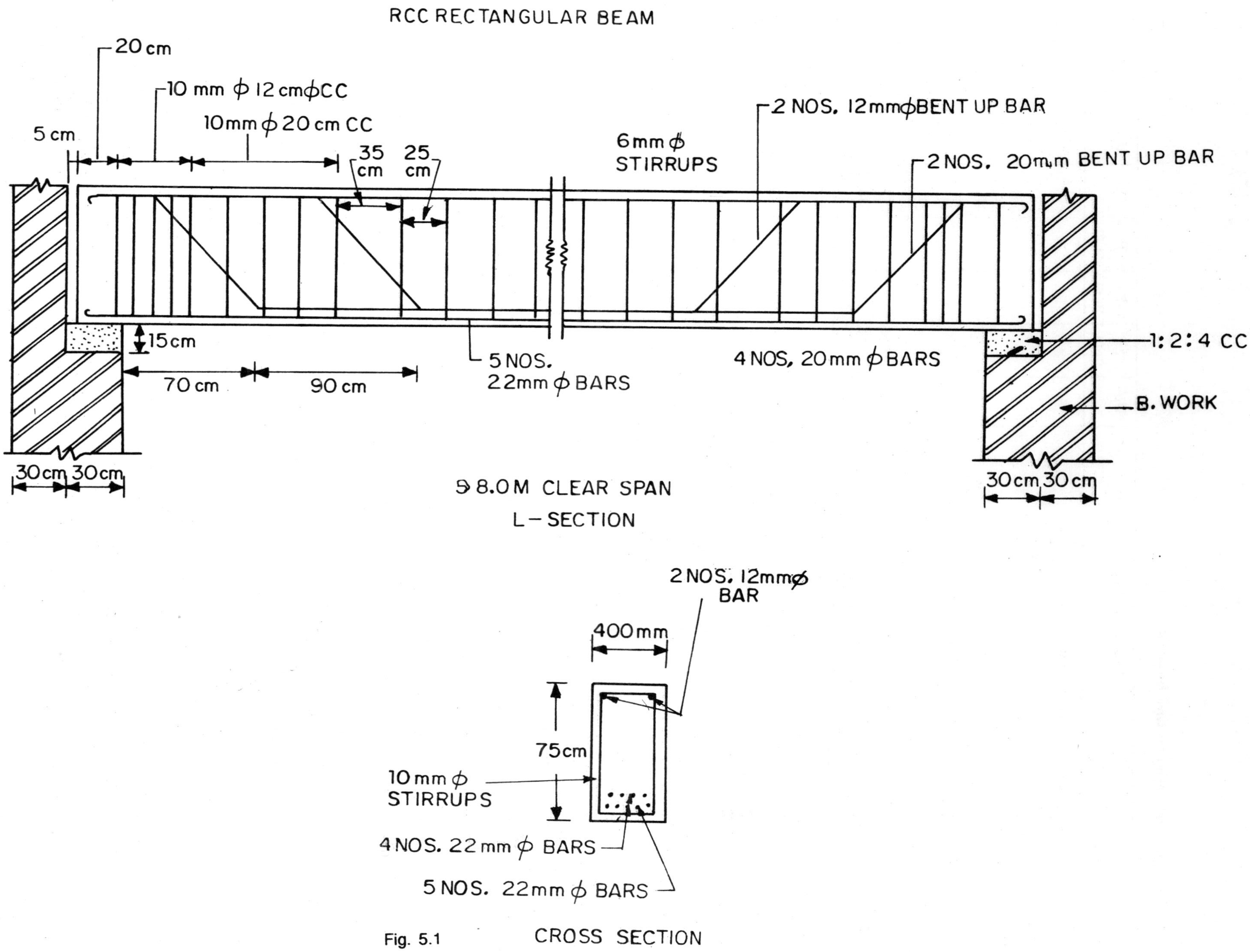
RCC RECTANGULAR BEAM
20 cm
10 mm ϕ 12 cmϕCC
10mm ϕ 20 cm CC
5 cm
35 cm
25 cm
6mm ϕ STIRRUPS
2 NOS. 12mmϕBENT UP BAR
2 NOS. 20mm BENT UP BAR
15cm
70 cm
90 cm
5 NOS. 22mm ϕ BARS
4 NOS, 20mm ϕ BARS
1:2:4 CC
B. WORK
30cm 30cm
30cm 30cm
8.0M CLEAR SPAN
L – SECTION
2 NOS. 12mmϕ BAR
400mm
75cm
10mm ϕ STIRRUPS
4 NOS. 22 mm ϕ BARS
5 NOS. 22mm ϕ BARS
CROSS SECTION

Fig. 5.1

Item No.	Description of Item	No.	L in m	B in m	H in m	Qty	Den.
	12 mm ϕ straight top bars	2	8.8	–	–	17.6	metre
	$L = 8.60 - \frac{2 \times 5}{100} + \frac{18 \times 12}{1000}$ = 8.716 ~ 8.8 metre						
	$\text{Wt.} = 17.6 \times \frac{\pi}{4} \times (1.2)^2 \times \frac{1}{100 \times 100} \times 7.5 \times 10$ = 0.15 Quintal						
3.	**Stirrups**						
	10 mm ϕ bars, $L = (65 \times 2 + 30 \times 2 + 24) \times 0.01 = 2.14$					2.14	metre
	At end 12 cm C. to C.	5 × 2	2.14			17.12	metre
	At end 20 cm C. to C.	4 × 2	2.14			38.52	metre
	$\text{Wt.} = \frac{\pi}{4} \times \frac{(1.0)^2}{100 \times 100} \times 38.52 \times 7.5 \times 10$ Quintal = 0.22 Quintal						
	Stirrups 6 mm ϕ						
	35 cm. C. to C.	2	2.14	–	–	4.28	metre
	At central remaining length of 25 cm C. to C.	22	2.14	–	–	47.08	metre
					Total	51.36	m
	No. of stirrups = $\frac{524}{25}$ + 1 = 21 + 1 = 22 Nos.						
	$\text{Wt.} = 53.36 \times \frac{\pi}{4} \times \frac{(0.6)^2}{100 \times 100} \times 7.5 \times 10$ Q = 0.11 Q.						
	Total wt. of steel = 2.83 Quintal						

Analysis of Rates

As in previous examples

(a) Rate of cement concrete 1 : 2 : 4 — Rs 670.00 per m^3

(b) Rate of Reinforcing bars, including bending and bending, etc. — Rs 9.12 per kg

Abstract of cost

(a)	Cement concrete 1 : 2 : 4 @ 670.00/m^3	1728.60
(b)	Reinforcing 2.83 Quintal @ 9.12/kg	2580.00
	Total	4308.60 say Rs 4310.00

Example 5.2 Estimation of a Channel. Length — 1 km.

Estimation of channel excavation as per Fig. 5.2 (a) and 5.2 (b).

Item No.	Description of Item	No.	Length m	Breadth m	Height m	Quantity m^3
1.	Excavation Work					
	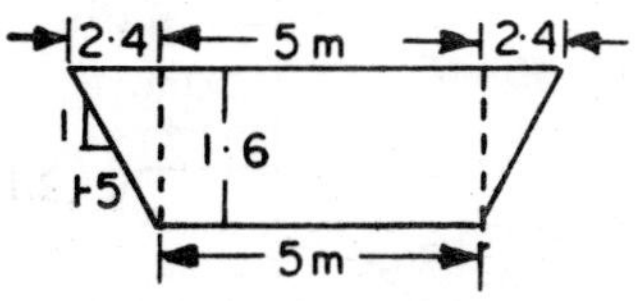 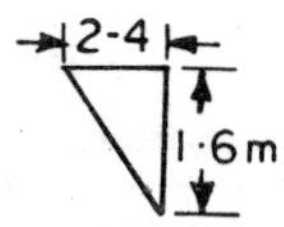	1	1000	5	1.6	8000
		2	½ × 1000	2.4	1.6	3840
					Total	11840
2.	Earth Filling					
	(a) Left bank	1	1000	3.0	0.95	2850
	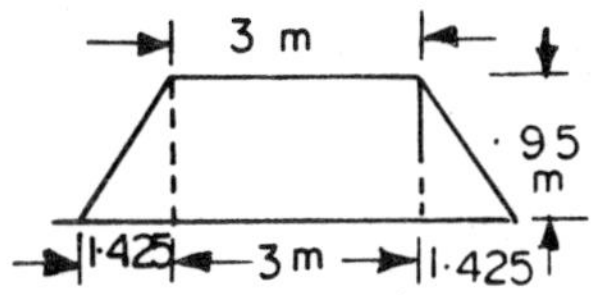 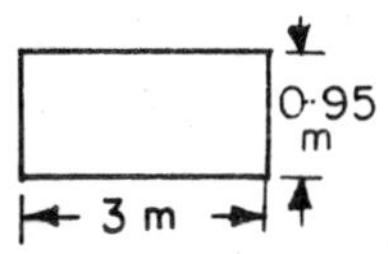 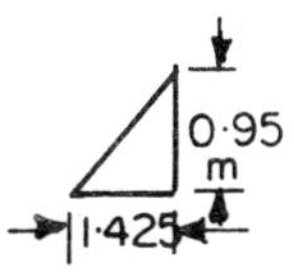	2	½ × 1000	1.425	0.95	1353.75
	(b) Right bank	1	1000	1.5	0.95	1425
	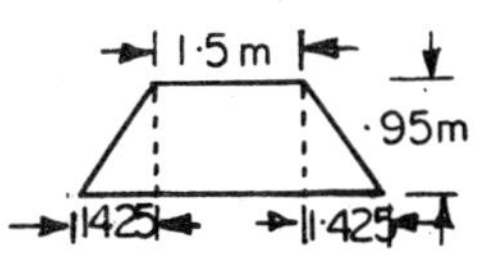 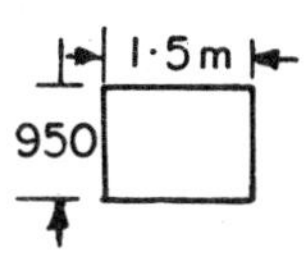 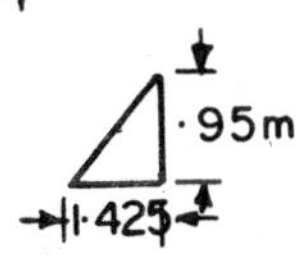	2	½ × 1000	1.425	0.95	135375
					Total	6982.5

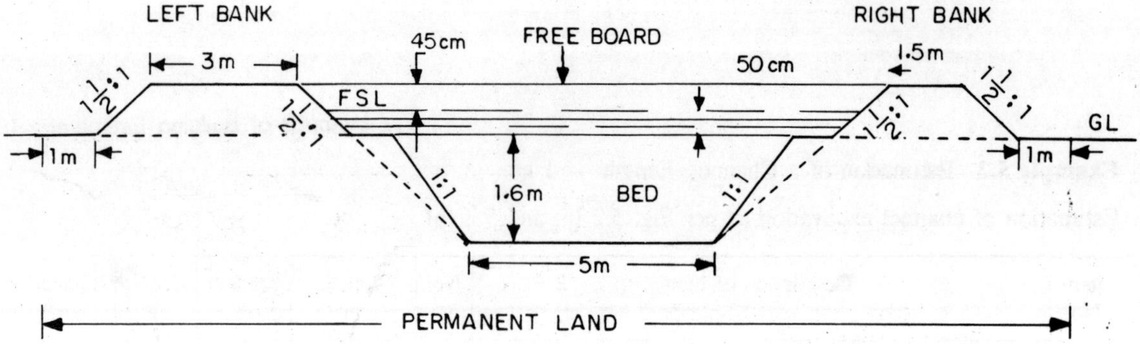

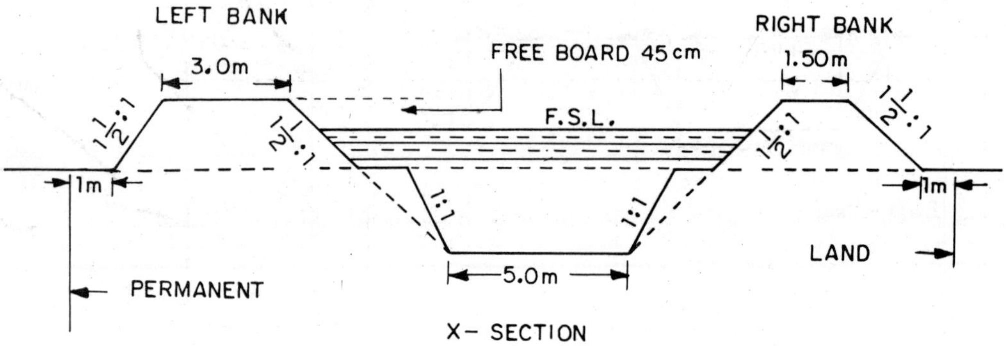

X- SECTION

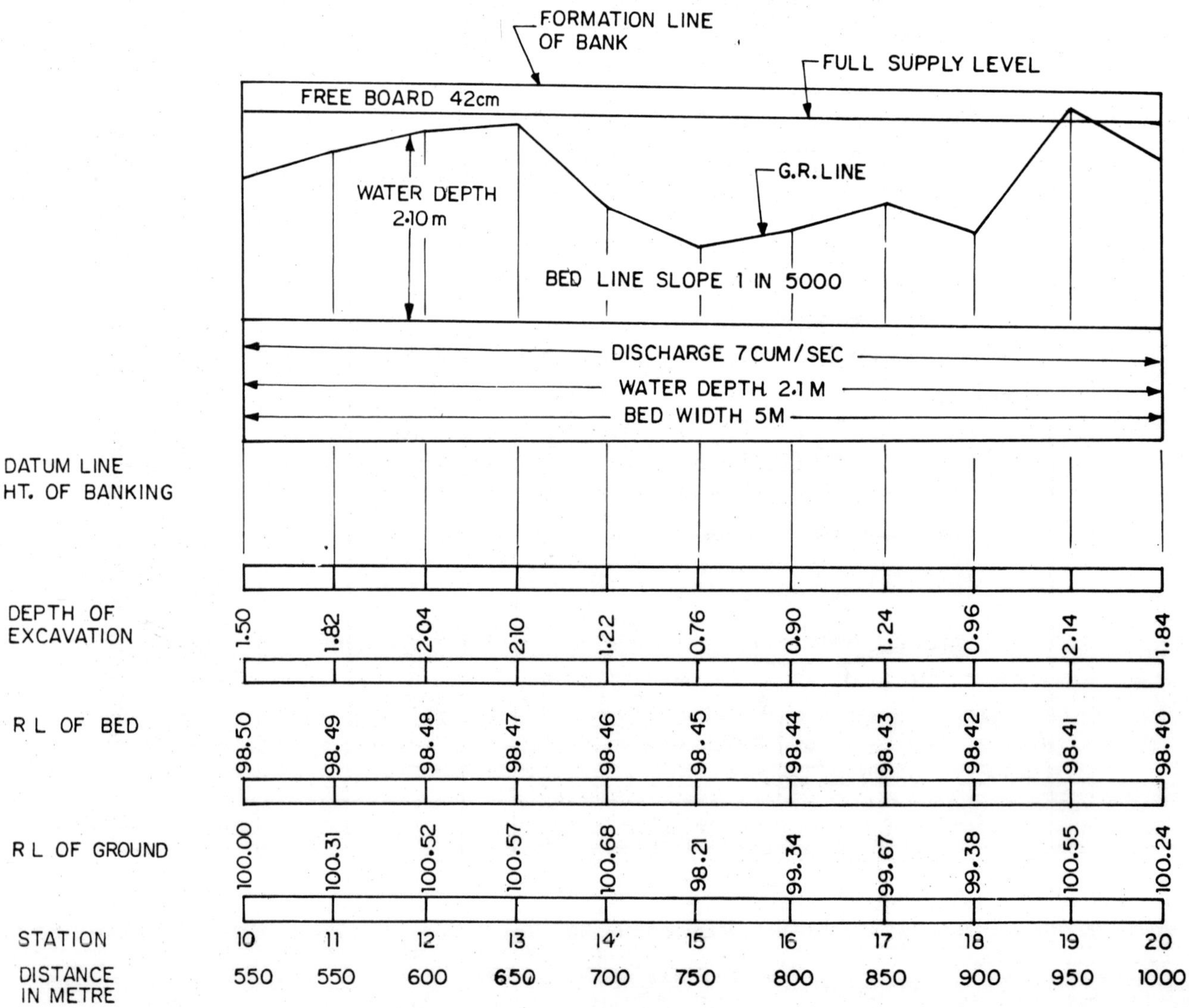

Fig. 5.2 (b)

Calculation of Quantities

Excavation in Digging in Channel and in Banking in Embankments

Excavation								Banking								
Station or Chaining	Depth of Digging	Mean Depth	Central Area	Side Area	Total Area	Length	Quantity	Height of Bank Above Bed	Height of Bank Above G.L.	Central Areas	Side Areas	Total Areas	Length	Quantity	Balance Quantity of Banks	Remark
		d	bd	s_d^2	$bd + s_d^2$	L	$Q_1 = (bd + s_d^2) L$	H	$H - d$	$(b_1 + b_2) L$	$2s_1L^2$	$(b_1 + b_2) L + 2S_1L^2$	L	Q_2	$Q_2 - Q_1$	
	m	m	m^2	m^2	m^2	m	m^3	m	m	m^2	m^2	– m^2	m	m^3	m^3	
1	2	3	4	5	6	7	8	9	10	11	12	13	14	15	16	17
10	1.50	—	—	—	—	—	—	—	—	—	—	—	—	—	—	Column 2 — Difference of R.L. of Ground and R.L. of Bed
11	1.82	1.66	8.30	2.756	11.056	50	552.80	2.55	0.89	4.005	2.376	6.381	50	319.05	+233.75	
12	2.04	1.93	9.65	3.725	13.375	50	668.75	2.55	0.62	2.790	1.153	3.943	50	197.15	+471.60	
13	2.10	2.07	10.35	4.285	14.635	50	731.75	2.55	0.48	2.160	0.691	2.851	50	147.55	+589.20	
14	1.22	1.66	8.30	2.756	11.056	50	552.80	2.55	0.89	4.005	2.376	6.381	50	315.05	+237.75	
15	0.76	0.99	4.85	0.980	5.930	50	206.50	2.55	1.56	1.020	7.302	14.321	50	116.05	–419.55	
16	0.90	0.83	4.15	0.689	4.839	50	241.95	2.55	1.72	7.140	8.875	16.615	50	830.75	–588.80	
17	1.24	1.07	5.30	1.145	6.495	50	324.75	2.55	1.48	6.660	6.577	12.237	50	661.85	–337.10	
18	0.96	1.10	5.50	1.210	6.710	50	335.50	2.55	1.45	6.525	6.308	12.833	50	641.50	–306.00	
19	2.14	1.55	7.75	2.403	10.153	50	507.65	2.55	1.00	4.500	3.000	7.500	50	375.00	+132.65	
20	1.84	1.99	19.95	3.960	13.910	50	695.50	2.55	0.56	2.520	0.941	3.461	50	173.05	+512.45	
						Total	4907.5				Total	87.523		4376.00	+534.95	

Note : Surplus excavated earth will be embanked as spoil bank near the channel to be utilised subsequently in repair work.

Quantity of earthwork in channel = 4907.5 cum.

Quantity of earthwork in bank (balance quantity required to be disposed of) = +534.95.

As per the analysis of rates in previous problems, rate of excavation is Rs 8.80/m^3.

Hence, cost of channel excavation = 8.80 × 4907.50 = 43,186.00.

Example 5.3 Estimate of Spiral Staircase (Refer Fig. 5.3)

Footing (foundation) — 1 : 2 : 4 R.C.C. work
Central Column and Railing — 1 : 2 : 4 R.C.C. work
Step of Stair — 1 : 1½ : 3 precast R.C.C. work
No of Steps — 12 Nos
Rise — 200 mm

Analysis of Rates

1. Earth work in Foundation Excavation
0.04 (10 A + B) = 0.04 (200 + 40) = 0.04 × 240 = Rs 9.60/m^3

2. Cement concrete 1 : 1½ : 3

a. Material

(i)	Cement bags 7.56 @ Rs 70.00 per bag	=	529.20
(ii)	Sand 0.4 m^3 @ Rs 125.00/m^3	=	50.00
(iii)	Coarse aggregate 0.8 m^3 @ 125.00/m^3	=	100.00
			679.20

b. Labour

$(2A + 8B + 2C + D/4) \times 0.4$ = Rs 116.80
= (80 + 60 + 40 + 12) × 0.4
= 292 × 0.4

Rs 796.00

Rs 796.00/m^3

3. Mild Steel 800 Rs/quintal

4. Bending building and cutting reinforcement =
$(3/4\ F + B) \times 2.25$ = (¾ × 40 + 20) × 2.25 = 50 × 2.25 = 112.50

Abstract of Cost

Item No.	Particulars of Items	Quantity	Units	Rate	Cost	Remark
1.	Excavation in foundation	0.3645	m^3	9.60	349.00	
2.	Cement concrete 1 : 2 : 4 in footing	0.2907	m^3	670.00	194.76	as evolved earlier.
3.	Railing	7.0	quintal	800.00	5600.00	
4.	Cement concrete 1 : 1½ : 3	0.63	m^3	796.00	502.11	
5.	Reinforcement steel	0.87	quintal	800.00	696.00	
					7341.87	
6.	Bending and cutting reinforcement	0.87	quintal	112.50	97.87	
					7439.74 say Rs 7450	

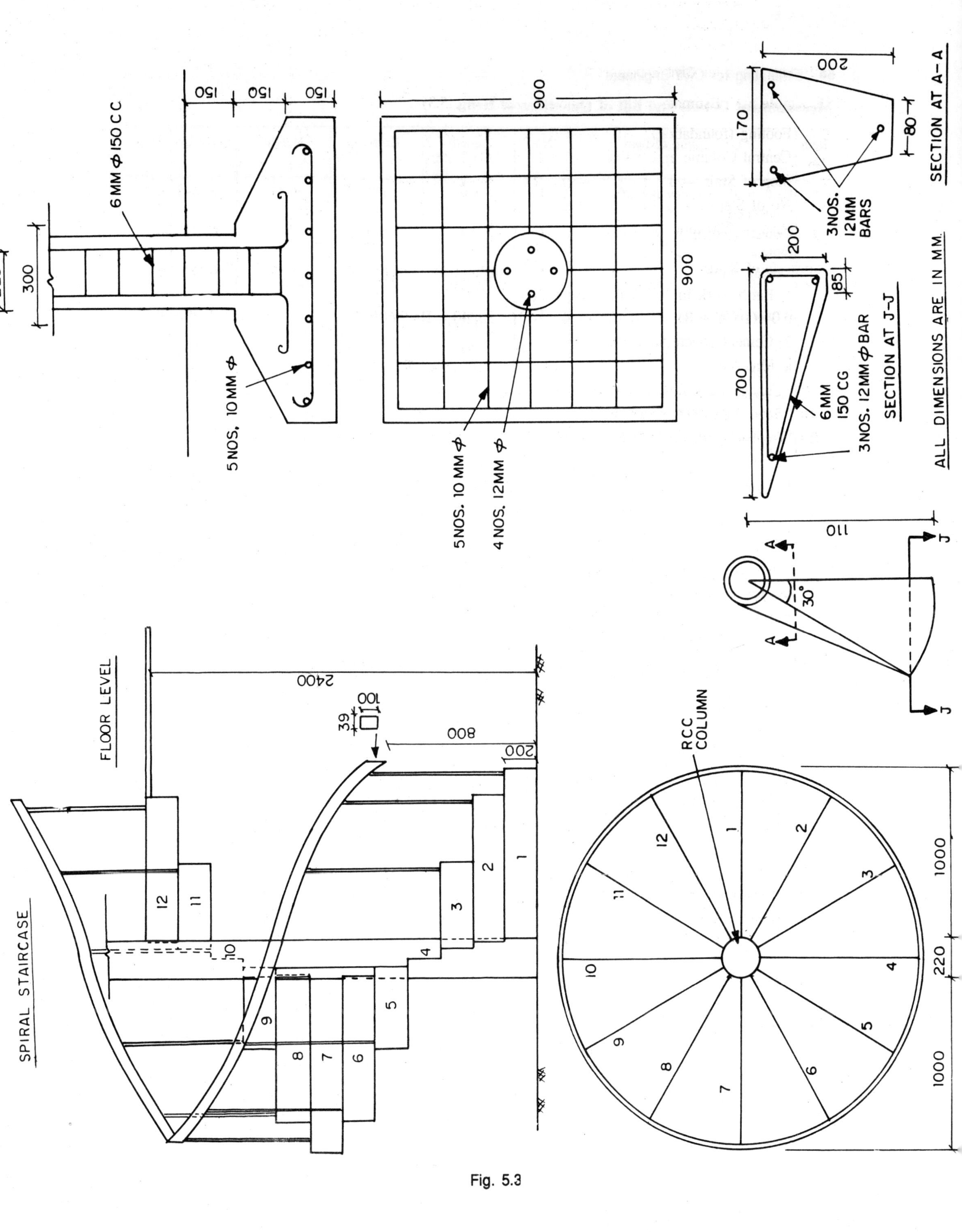
SPIRAL STAIRCASE
FLOOR LEVEL
2400
800
200
39
100
RCC COLUMN
1000
220
1000
6 MM Φ 150 C C
5 NOS. 10 MM Φ
220
300
150
150
150
900
900
5 NOS. 10 MM Φ
4 NOS. 12MM Φ
110
30°
A
A
J
J
700
200
85
6 MM 150 CG
3 NOS. 12MM Φ BAR
SECTION AT J–J
200
170
80
3NOS. 12MM BARS
SECTION AT A–A
ALL DIMENSIONS ARE IN MM.

Fig. 5.3

Measurement Details and Bill of Quantities of Items

Item No.	Description of Item	No.	Length	Breadth	Height	Quantity	Unit
1.	Earth work in excavation of foundation	1	0.9	0.9	0.45	0.3645	m^3
2.	Cement concrete 1 : 2 : 4 in footing	1	0.4	0.9	0.19	0.122	m^3
	Sloping portion	1	$\frac{(0.30+0.9)}{2}$	$\frac{(0.3+0.9)}{2}$	0.15	0.054	m^3
	Central column	1	$\frac{\pi}{4}(0.22)^2$ $\times 315$			0.1197	m^3
	Railing Sectional area (vertical) = (0.08 × 0.1) + 10.04 × 0.8) = 0.04 m^2 Plan length = 1⅙ perimeter = 1.167 × π × 2.32 = 8.139 m.						
				0.04	8.139	0.328	m^3
					Total	0.106	m^3
					or 0.106 × 70 quintals = 7 Quintals		
3.	1 : 1½ : 3, R.C.C. work (Precast) in steps including moulds. 2. Fixing in position for one step. Cross-Section area at A-A = $\frac{(0.17-0.08)}{2} \times 0.2 = 0.025\ m^2$ Section area at J-J is $\frac{(0.70+0.08)}{2} \times 0.2 = 0.078\ m^2$						
	Quantity in steps from point A to outer periphery	12	$\frac{0.025+0.078}{2}$ $\times 102$			0.63	m^3
	Central Portion around column = πdth	12	0.19	0.03	0.2	0.05	m^3
					Total	0.063	m^3
4.	M.S. Reinforcement including bending and placing in position						
	12 mm dia bar	14 × 3	1			42.0	m.
	10 mm dia bar in footing	5 × 2	0.9			9.0	m
	Column						
	6 mm dia bar step stirrups of mean length in railing	14 × 6	1.0			84.0	m
		40	1.0			40.0	m
		4	9.0			36.0	m
						160.0	m
						0.22	kg/m
					Total	87.77	kg

Example 5.4 Estimation of RCC Column (Refer Fig. 5.4).

Analysis of Rates

1. Steel @ Rs 800.00 per quintal
2. Earth work in foundation

= 0.04 (10 A + B)

= 0.04 (200 + 40) = 0.04 × 240 = Rs 9.60/m^3

3. Cement concrete 1 : 4 : 8 in foundation

a. Material

Cement bags 3.25 @ 70/- per bag	=	227.50
Sand 0.45 m^3 @ Rs 5.00/m^3	=	56.00
Coarse aggregate 0.90 m^3 @ Rs 125.00/m^3	=	112.00
		395.50

b. Labour

$(2A + 8B + 2C + D/4) \times 0.4$

or 0.4 × (80 + 160 + 40 + 12) = 292 × 0.4 = Rs 116.80 per unit

Total = 395.50 + 116.80

= 512.30/m^3

4. Cement concrete 1 : 2 : 4 in column and footing

a. Material

Cement 5.76 bags @ 70.00 per bag	= 403.20
Sand 0.40 m^3 @ 125.00 per m^3	= 50.00
Coarse aggregate 0.80 m^3 @ 125.00 per m^3	= 100.00
	553.20

b. Labour

Rs 116.80 per m^3 as above

Total 553.20 + 116.80 = Rs 670.00/m^3

5. Bending, binding and cutting of bars

(3/4 F + B) × 2.25 = (3/4 × 40 × 20) × 2.25 = 112.50 per quintal

Abstract of Cost

Item No.	Particulars of Item	Qty	Units	Rate	Cost
1.	Earth work in excavation of foundation	23.88	m^3	9.60	230.00
2.	Cement concrete 1 : 4 : 8 in base	0.49	m^3	512.30	251.00
3.	Cement concrete 1 : 2 : 4	7.674	m^3	670.00	5,120.00
4.	Mild steel	331	kg	8.00	2,648.00
					8,249.00
	Shuttering etc.	L.S.			500.00
					8749.00
	Add contingency @ 5%				436.45
					9286.45
					say 9,300.00

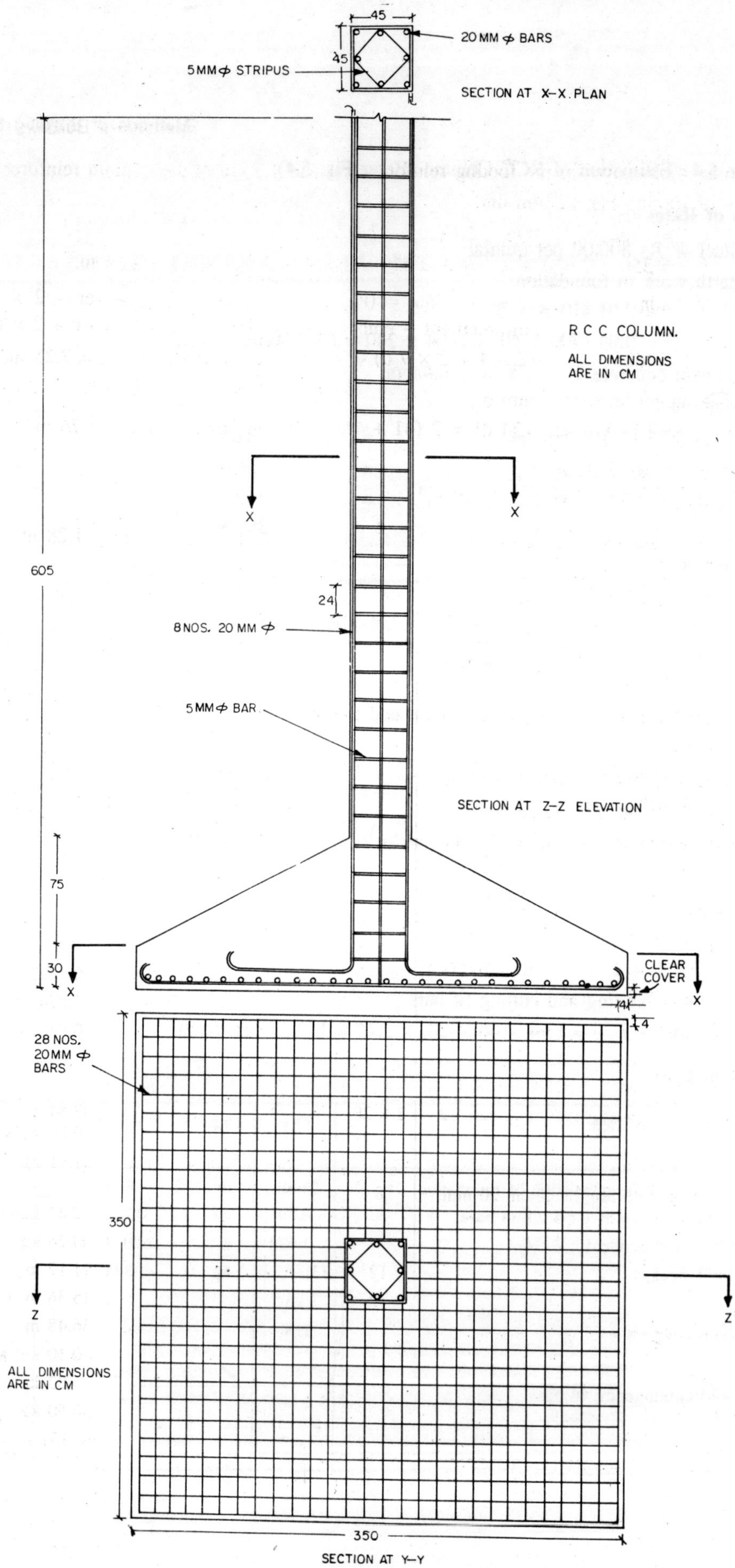
45
45
20 MM ϕ BARS
5 MM ϕ STRIPUS
SECTION AT X–X. PLAN
R C C COLUMN.
ALL DIMENSIONS ARE IN CM
605
24
8 NOS. 20 MM ϕ
5 MM ϕ BAR
SECTION AT Z–Z ELEVATION
75
30
X
X
CLEAR COVER
4
4
28 NOS. 20 MM ϕ BARS
350
Z
Z
ALL DIMENSIONS ARE IN CM
350
SECTION AT Y–Y

Fig. 5.4

Assuming 4 cm cover in the footing reinforcement and 1.5 cm in the column reinforcement.

Length of footing bar 12 mm dia

= width of footing – 2 Cores + 2 Hooks

= 350 – 2 × 4 + 2 × 9 d = 350 – 2 × 4 + 2 × 9 × 1.2 = 3.64 m.

Length of main column bar 20 mm dia = (Length of column – base cover – 2 × dia of base reinforcement + embedded horizontal portion in base + 2 × hook length)

= (605 – 2 × 1.2 – 4 + 2 × 9 d) + 90 (embedded) = 724.6 cm = 7.25 metre

Length of larger stirrups 5 mm ϕ

= [2 (A + B) + 24 d] = 2 (41 + 41) + 24 × 0.5 = 176 cm. = 1.76 m

No of strips $= \dfrac{605 - 4 - 2 \times 12}{24} = 25$ Nos.

Length of smallest stirrups = $[2(25.5\sqrt{2} + 25.5\sqrt{2}) + 24 \times 5] = 127.96$ cm $\simeq$ 1.28 m

Item No.	Description of Item	No.	Length	Breadth	Height	Quantity	Total
1.	Earth work in excavation in foundation	1	3.5	3.5	1.95	23.8875 m^3	
2.	Cement concrete 1 : 4 : 8 Base	1	3.5	3.5	0.04	0.49 m^3	
3.	R. C. C. Work 1 : 2 : 4 in footing including steel						
	Bottom square portion	1	3.5	3.5	0.26	3.185 m^3	
	Trapezoidal Portion	1					
	$\frac{H}{6}(A_1 + A_2 + 4\,Am)^2$						
	$= \frac{0.75}{6}\left[3.5^2 + 0.45^2 + 4\left(\frac{3.5 + 0.45}{2}\right)^2\right]$						
	= 3.507 m^3						
	Open Square Portion	1	0.45	0.45	0.9	0.18225 m^3	
					Total =	6.874 m^3	
4.	R.C.C. Work 1 : 2 : 4 in column above ground level	1	0.45	0.45	4.1	0.83025 m^3	
	Steel Reinforcement						
1.	M.S. Work reinforcement footing bars 12 mm dia	2 × 28	3.64 m			203.84 m 0.89 kg/m	
					Total	181.41 kg	
2.	Column with main reinforcement 20 mm dia	8	7.25			58.0 m 2.47 kg/m	
					Total	143.26 kg	
3.	5 mm dia stirrups	12	1.76			21.12 m	
		12	1.28			15.36 m	
					Total	36.48 m	
						0.10 kg/m	
					Total	5.83 kg	
	Total Wt of Reinforcement					330.50 kg say 331 kg.	

Analysis of Rates

1. Earth work in excavation 0.04 (10 A + B) = 0.04 (200 + 20) = 0.04 × 220 = Rs 8.80/m^3

2. Cement Concrete 1 : 4 : 8

a. Material

Cement bags 3.25 @ Rs 70.00 per bag	=	227.50
Sand 0.45 m^3 @ Rs 125.00 per m^3	=	56.25
Coarse aggregate 0.90 m^3 @ Rs 125.00 per m^3	=	112.50
		396.25

b. Labour

(2 A + 8 B + 2 C + D/4) × 0.4

(80 + 160 + 40 + 12) × 0.4

= 292 × 0.4 = 116.80

Total = Rs 513.05/m^3

3. Brick work 1 : 6

Cement mortar

a. Material

Cement bags 1.35 @ Rs 70.00 per bag	=	94.50
Sand 0.275 m^3 @ Rs 125.00/m^3	=	34.37
Bricks 460 @ Rs 500 per 1000	=	230.00
		Rs 358.87/m^3

4. Earth filling — Excavated earth shall be utilised for earth filling. Only compaction charges will have to be paid.

= 0.4 A = 0.04 × 20 = 0.80/m^3

5. Sand filling rates are the same as obtaining sand i.e. Rs 125.00 per m^3

6. Broken Ballast filling -Rs 125.00/ m^3

7. Cement Concrete 1 : 2 : 4

Cement bag 576 @ Rs 70.00 per bag	=	403.20
Sand 0.40 m^3 @ Rs 125.00 per m^3	=	50.00
Coarse 0.80 m^3 @ Rs 125.00 per m^3	=	100.00
		553.20
Labour		116.80
Total		570.00

Reinforcement

Cost Rs 800.00 per quintal

Bending, Binding and cutting etc. = (3/4 F + B) × 2.25

= (3/4 × 40 + 20) × 2.25

= 50 × 2.25

= 112.50

Total = Rs 912.50 per quintal

Example 5.5 Estimation of Grain-Silo (Refer Fig. 5.5).

Item No.	Description of Item	Size	Height	Calculation	Quantity	Detail
1.	Earth Excavation	$r = 1.75$	0.53	$\frac{\pi}{4}(1.75)^2 \times 0.53$	1.2747 m^3	
2.	C.C. in Foundation	$r_1 = 1.75$ $r_2 = 0.75$	0.23	$\frac{\pi}{4} \times (1.75)^2 \times 0.23 - \frac{\pi}{4}(.75)^2 \times 0.23$	0.452 m^3	
3.	Brickwork in Foundation	$r_1 = 1.48$ $r_2 = 1.02$	0.3	$\frac{\pi}{4}(1.48^2 - 1.02^2) \times 0.3$	0.575 m^3	
4.	Brickwork above Ground Level	$r_1 = 1.48$ $r_2 = 1.02$	0.3	$\frac{\pi}{4}(1.48^2 - 1.02^2) \times 0.3$	0.575 m^3	
5.	Earth Filling	$r = 1.02$	0.28	$\frac{1}{2} \times \frac{\pi}{4}(1.02)^2 \times 0.24$	0.098 m^3	
	10% is added due to slope		+10%	–	0.0093 m^3	
				Total =	0.10336 m^3	
6.	Sand Filling	$r = 1.02$	0.15	$\frac{\pi}{4}(1.02)^2 \times 0.15$	0.122569 m^3	
	10% is added due to slope		10%	–	0.0122569 m^3	
				Total =	0.1348259 m^3	
7.	Broken Stone	$r = 1.02$	0.075	$\frac{\pi}{4}(1.02)^2 \times 0.75$	0.06178461 m^3	
	10% is added due to slope		10%		0.00612846 m^3	
				Total =	0.06741306 m^3	
8.	Cement Concrete Finish	1.02	0.011	$\frac{\pi}{4}(1.02)^2 \times 0.01$	0.008988	
	10% is added due to slope		10%		0.000898	
				Total =	0.009886 m^3	
9.	Outlet Pipe Precast Grainspout 15 cm Dia M.S. Pipe with Locking Arrangement	Dia 150 mm	Thickness 50 mm	–	–	
10.	R.C.C. Ring	$r_1 = 175$ $r = 1.2$	1.9	$\frac{\pi}{4}(1.25)^2 \times 1.9 - \frac{\pi}{4}(1.2)^2 \times 1.9$	0.1828 m^3	
11.	Concrete Cover					
		$r_1 = 0.54$ $r_2 = 40$	0.07	$\frac{\pi}{4} 1.54^2 - 0.070^2 \times 0.07$	0.0038 m^3	
		$r_1 = 1.3$ $r_2 = 1.23$	0.75	$\frac{\pi}{4}(1.3^2 - 1.23^2) \times 0.075$	0.010392 m^3	
		$r_1 = 1.35$ $r_2 = 1.2$	$h = 0.02$	$\frac{1}{3}\pi \times [(1.35)^2 - (1.25)^2] \times 0.02$		

Continued

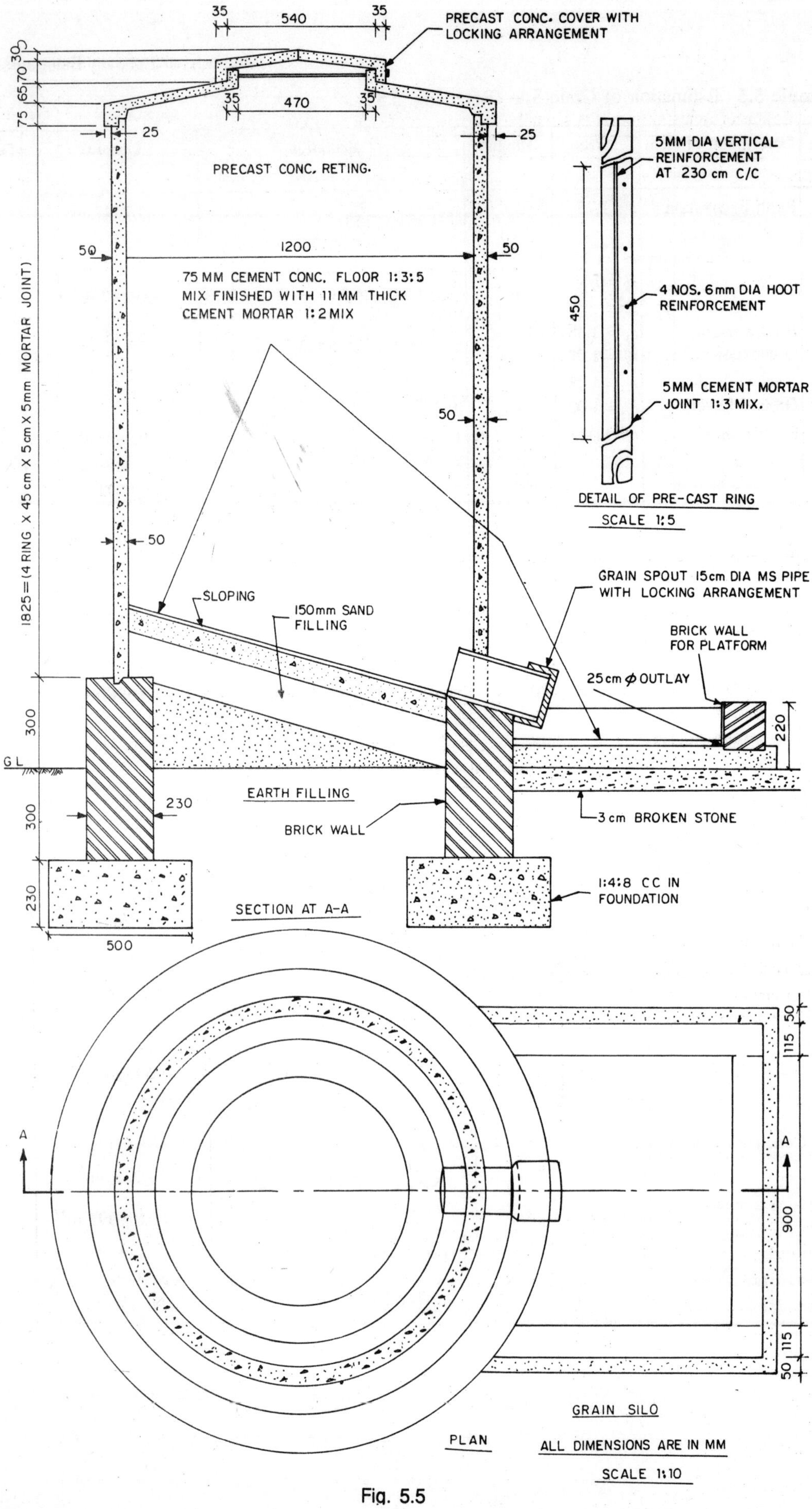

35
540
35
PRECAST CONC. COVER WITH LOCKING ARRANGEMENT
30
70
65
75
35
470
35
25
25
PRECAST CONC. RETING.
5MM DIA VERTICAL REINFORCEMENT AT 230 cm C/C
50
1200
50
75 MM CEMENT CONC. FLOOR 1:3:5 MIX FINISHED WITH 11 MM THICK CEMENT MORTAR 1:2 MIX
450
4 NOS. 6mm DIA HOOT REINFORCEMENT
5MM CEMENT MORTAR JOINT 1:3 MIX.
50
1825 = (4 RING X 45 cm X 5 cm X 5mm MORTAR JOINT)
DETAIL OF PRE-CAST RING
SCALE 1:5
50
GRAIN SPOUT 15cm DIA MS PIPE WITH LOCKING ARRANGEMENT
SLOPING
150mm SAND FILLING
BRICK WALL FOR PLATFORM
25cm ϕ OUTLAY
300
220
GL
EARTH FILLING
230
300
BRICK WALL
3 cm BROKEN STONE
1:4:8 CC IN FOUNDATION
230
SECTION AT A-A
500
50
115
A
A
900
115
50
GRAIN SILO
PLAN
ALL DIMENSIONS ARE IN MM
SCALE 1:10

Fig. 5.5

Item No.	Description of Item	Size	Height	Calculation	Quantity	Detail
		$r_1 = 1$		$\frac{1}{3}\pi\,[(.51)^2 - (.44)^2] \times 0.105$		
			Total =	0.00801 – 0.007312 = 0.007278 m^3		
			Total =	0.0869799 m^3		
		$r_1 = \frac{0.69}{2}$ $r_2 = \frac{0.54}{2}$	$h_1 = 0.07$ $h_2 =$ 0.035	$\frac{1}{3}\pi r_1^2 h_1 - \frac{1}{3}\pi r_2^2 h_2$	0.0004147 m^3	
	Cylindrical portion	$r_1 = \frac{0.61}{2}$ $r_2 = \frac{0.54}{2}$	$h = 0.07$	$\pi (r_1^2 - r_2^2)\, h$	0.00442 m^3	
				Total =	0.00857 m^3	
	Reinforcement					
1.	5 mm φ vertical reinforcement 230 C.C.	$r =$ 0.625 cm	Circum-ference = $2\pi r = 2 \times \pi \times 62.5$		392.699 cm	
				No. of bars = $\frac{392.699}{230}$	1.7 bars ≃ 2 bars	
				Total Hoopbar =	4 × 2 = 8 bars	
				Quantity = 8 Length = 0.45 m Total length of bar = 3.6 m		
	Hoop reinforcement 4 Nos 6 mm dia length of 1 bar	circum-ference + 24 d = 407.099				
	Total length of bar			4 × 4 × 407.009	6513.584 cm 65.13 m ≃ 66 m	
	For platform					
1.	Earth excavation	0.875	1.23	0.03	0.0323 m^3	
2.	Broken stone	0.875	1.23	0.03	0.0323 m^3	
3.	Brick work (2)	0.71	0.22′	0.115	0.036 m^3	
		1.13	0.22	0.115	0.0286 m^3	
				Total –	0.0646 m^3	
4.	Plain C.C. in platform	0.825	1.13	0.07	0.065 m^3	
5.	Platform floor finish	0.71	0.9	0.075	0.047 m^3	
	Total quantities					
	Total Earth Excavation				1.307 m^3	
	Total Brick Work				0.0646 m^3	
	Total Broken Stone				0.05925 m^3	
	Plain C.C.				0.5173 m^3	
	Floor Finish				0.15385 m^3	

Example 5.6 Estimation of Plate Girder (Refer Fig. 5.6).

Item No.	Description of Item	No.	Length	Breadth	Height	Quantity	Total
1.	Web Plate 12 mm thick	2	1.5	1.0		3.0 m^2	
		2	1.5	1.0		3.0 m^2	
		1	12.0	1.0		12.0 m^2	
					Total	18 m^2	
						94.2 kg/m^2	
					Total wt	1695.6 kg	
2.	20 mm thick Flange Plate	2	1.5	0.5		1.5 m^2	
						157.0 kg/m^2	
					Total wt	235.5 kg	
3.	32 mm thick Flange Plate	2	1.5	0.5	–	1.5 m^2	
						251.2 kg/m^2	
					Total wt	376.8 kg	
4.	45 mm thick Flange Plate	2	12.0	0.5		12 m^2	
						352.2 kg/m^2	
					Total wt	2660.1 kg	
5.	Stiffener plates 10 mm thick	12	1.0	0.144		1.728 m^2	
						78.5 kg/m^2	
					Total wt	135.6 kg	
	Total Weight of Steel					2795.7 kg	
	Rate of Structural Steel Rs 8.00 per kg						
	Cost of the plate girder steel					22365.6	
	Add 5% for welding and contigency					1118.28	
					Total Cost	23483.88	
						Say Rs 24,000.00	

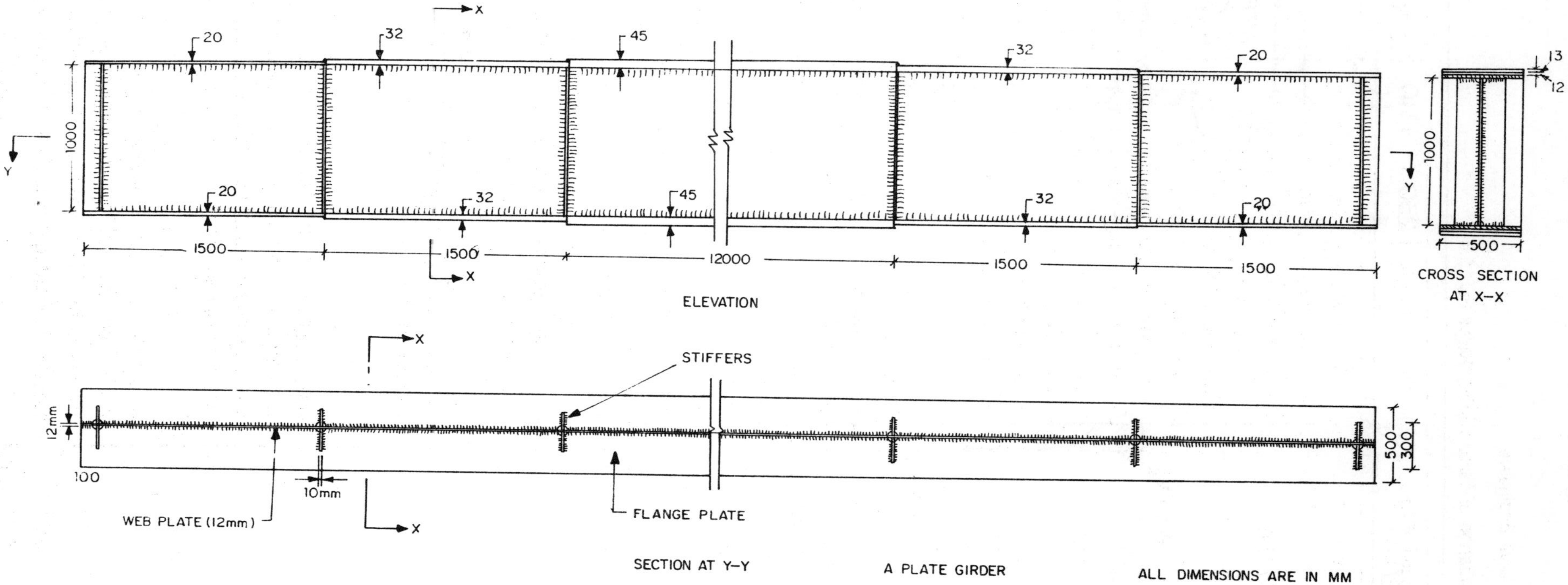

20
32
45
32
20
13
12
1000
20
32
45
32
20
1000
1500
1500
12000
1500
1500
500
ELEVATION
CROSS SECTION
AT X-X
STIFFERS
12mm
500
300
100
10mm
WEB PLATE (12mm)
FLANGE PLATE
SECTION AT Y-Y
A PLATE GIRDER
ALL DIMENSIONS ARE IN MM

Fig. 5.6

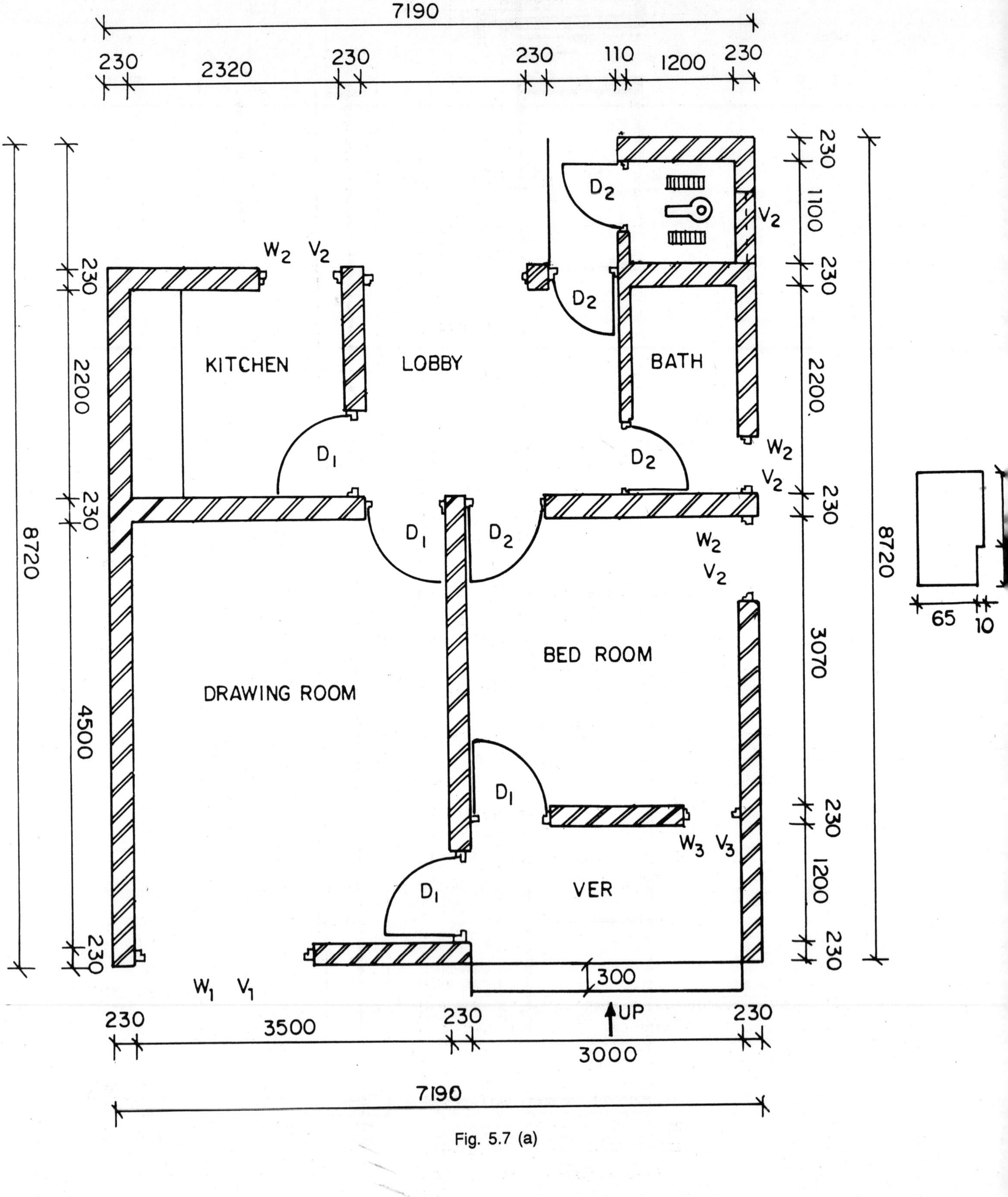
7190
230
2320
230
230
110
1200
230
KITCHEN
LOBBY
BATH
DRAWING ROOM
BED ROOM
VER
D1
D2
W1 V1
W2 V2
W3 V3
UP
300
3500
3000
4500
2200
3070
1100
8720
65
10

Fig. 5.7 (a)

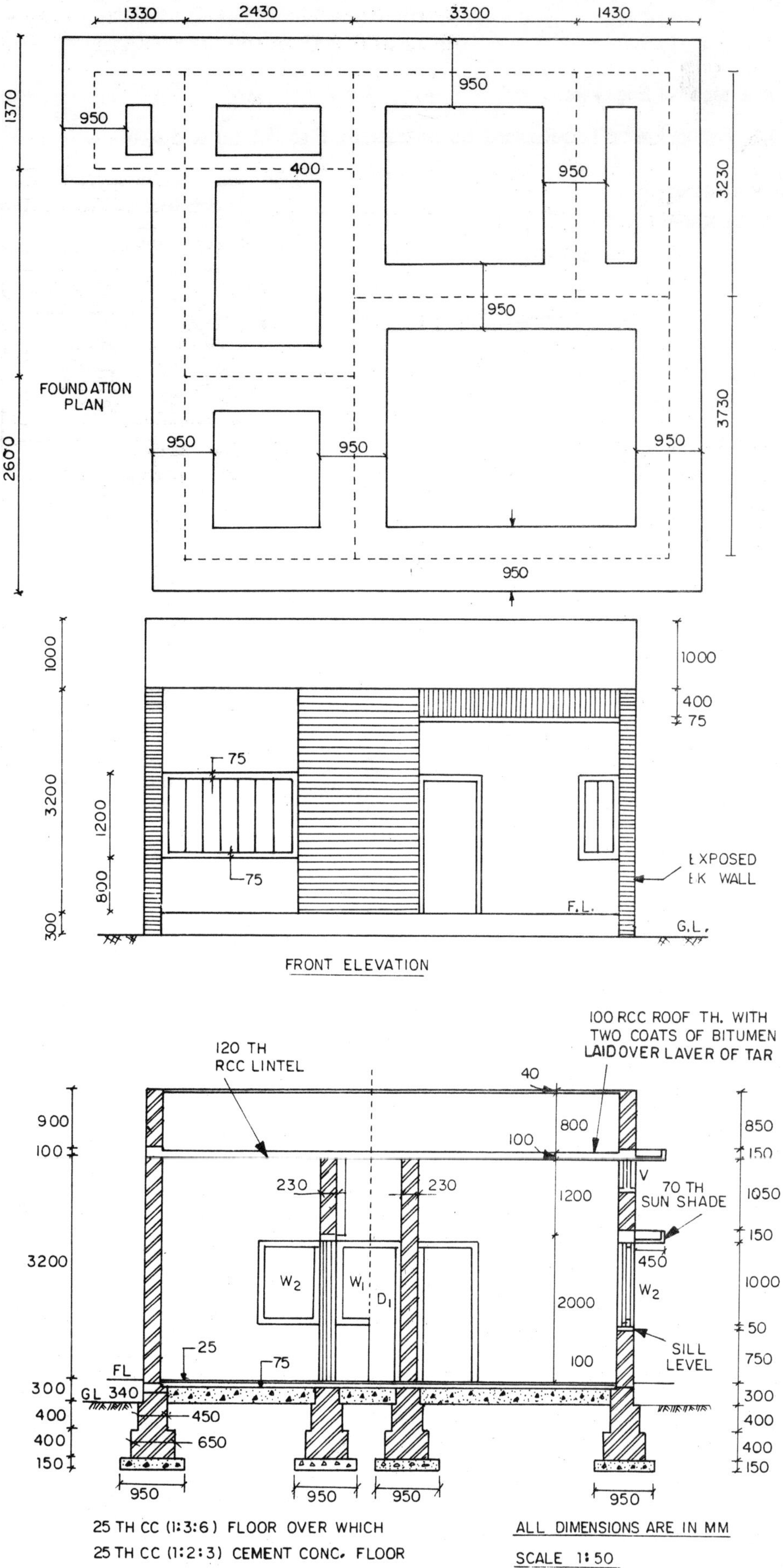

1330
2430
3300
1430
1370
950
400
950
3230
950
FOUNDATION PLAN
2600
950
950
950
3730
950
1000
3200
1200
800
300
75
75
1000
400
75
EXPOSED BK WALL
F.L.
G.L.
FRONT ELEVATION
100 RCC ROOF TH. WITH TWO COATS OF BITUMEN LAID OVER LAYER OF TAR
120 TH RCC LINTEL
40
900
100
800
100
850
150
230
230
1200
70 TH SUN SHADE
1050
150
450
3200
W2
W1
D1
2000
W2
1000
50
25
75
100
SILL LEVEL
750
FL
300
GL 340
300
400
450
400
400
650
400
150
150
950
950
950
950
25 TH CC (1:3:6) FLOOR OVER WHICH
25 TH CC (1:2:3) CEMENT CONC. FLOOR
ALL DIMENSIONS ARE IN MM
SCALE 1:50

Fig. 5.7 (b)

Example 5.7 Estimation of Two-Roomed Building [Refer Figs. 5.7 (a) & 5.7 (b)].

Lengths of Various Walls

1. Drawing Room
   Long wall - 4.73 m - (2)
   Short wall - 3.73 m - (2)
2. Right Hand Side Bed Room
   Long Wall - 3.3 m - (1)
   Short wall - 3.23 m - (2)
3. Front Verandah
   Long wall - 3.23 m - (1)
   Side wall - 1.43 m - (1)
4. Kitchen
   Long wall - 2.6 m - (1)
   Short wall - 2.43 m - (2)
5. Right hand side of kitchen
   Short 2.99 m - (1)
6. Latrine
   Short wall - 1.37 m - (2)
   Long wall - 1.33 m - 2 (1 + 1)
7. Bath
   Long wall - 2.43 m - 2 (1 + 1)

Schedule of Doors & Windows

D/W	Width (mm)	Height (mm)
D_1	900	2000
D_2	750	2000
W_1	2000	1200
W_2	900	1200
W_3	600	1200

D — Door
W — Window

Details of Measurement and Quantities

Item No.	Particulars of Item	No.	Length	Breadth	Depth	Quantity	Explanation
A.	Earthwork in Foundation						
1.	Drawing Room						
	Long wall	2	5.68	0.95	0.95	10.252 m^3	4.73 + 0.95
	Short wall	2	2.78	0.95	0.95	5.0179 m^3	3.73 – 0.95
2.	Front Verandah						
	Long wall	1	4.8	0.95	0.95	3.77245 m^3	3.231 – 0.95
	Short wall	1	0.48	0.95	0.95	0.4332 m^3	1.43 – 0.95
3.	Right Hand Side Bed Room						
	Long wall	1	4.25	0.95	0.95	3.835625	3.3 + 0.95
	Short wall	2	2.28	0.95	0.95	4.1154	3.23 – 0.95
4.	Kitchen						
	Long wall	1	3.55	0.95	0.95	3.2003875	2.6 + 0.95
	Short wall	2	1.48	0.95	0.95	2.6714	2.43 – 0.95
5.	Right hand side of kitchen						
	Long wall	–	–	–	–	–	–
	Short wall	1	2.895	0.95	0.95	2.74	2.99 – 0.95

Continued

Item No.	Particulars of Item	No.	Length	Breadth	Depth	Quantity	Explanation
6.	Latrine						
	Long wall	1	2.28	0.4	0.95	0.8664	1.33 + 0.95
		1	2.28	0.95	0.95	2.0577	
	Short wall	2	0.42	0.95	0.95	0.4	1.37 – 0.95
7.	Bath						
	Long wall	2	1.48	0.4	0.95	0.5624 × 2	2.43 – 0.95
	Short wall 1	1.48	0.95	0.95	1.3357 = 1.8981 m^3		
					Total	40.19815 m^3 $\simeq$ 41 m^3 + 1.3357 m^3 $\simeq$ 42 m^3	
B.	Lime Concrete in Foundation Bed (in m^3)						
1.	Drawing Hall						
	Long wall	2	5.68	0.95	0.15	1.6188	Length and width being the same as the excavation
	Short wall	2	2.78	0.95	0.15	0.7923	
2.	Front Verandah						
	Long wall	1	4.18	0.95	0.15	0.59565	
	Short wall	1	0.48	0.95	0.15	0.0684	
3.	Right hand side bed room						
	Long wall	1	4.25	0.95	0.15	0.605052	
	Short wall	2	2.28	0.95	0.15	0.6498	
4.	Kitchen						
	Short wall	2	1.48	0.95	0.15	0.42	
	Long wall	1	3.55	0.95	0.15	0.505875	
5.	R.H.S. Kitchen						
	Long wall	–	–	–	–	–	
	Short wall	1	2.315	0.95	0.15	0.32988	
6.	Latrine						
	Long wall	1	2.78	0.4	0.15	0.1368	
		1	2.28	0.95	0.15	0.3249	
	Short wall	2	0.42	0.95	0.15	0.1197	
7.	Bath						
	Long wall	1	1.48	0.4	0.15	0.0888	
		1	1.48	0.95	0.15	0.2109	
					Total	6.25773 m^3	
C.	Brick Work in Foundation upto Plinth (in m^3)						
1.	Drawing Room						
	1st footing						
	Long wall	1	5.53	0.65	0.4	1.4378	4.73 + 0.95
		1	5.38	0.65	0.4	1.3988	4.73 + 0.65

Continued

Item No.	Particulars of Item	No.	Length	Breadth	Depth	Quantity	Explanation
	Short wall	2	3.43	0.4	0.65	1.58	1.7836 3.73 – 0.3
	2nd footing						
	Long wall	1	5.43	0.45	0.4	0.9774	4.73 + 95/2 + 54/2
		1	5.18	0.45	0.4	0.9324	4.73 + 0.45
	Short wall	2	3.23	0.45	0.4	0.5814 × 2 = 1.1628	3.73 – 0.5
	Plinth wall above footing						
	Long wall	2	5.32	0.34	0.2	0.36176	4.73 + 9½ + 0.115
			5.015	0.34	0.2	0.34102	473 + 0.115 + 0.1
	Short wall	2	3.335	0.34	0.2	0.45356	373 – 0.45/2 – 0.17
2.	Front verandah						
	1st footing						
	Long wall	1	2.58	0.65	0.4	0.5814	3.23 – 4.5/2 + 0.4/2
	Short wall	1	0.78	0.65	0.4	0.2028	1.43 – 0.65
	2nd Footing						
	Long wall	1	3.23	0.45	0.4	0.5814	3.23 – 4.5/2 + 0.4/2
	Short wall	1	0.98	0.45	0.4	0.1764	1.43 – 0.45
	Plinth wall above footy						
	Long wall	1	3.285	0.34	0.2	0.22338	3.23 – 0.34/2 + 0.45
	Short wall	1	1.035	0.34	0.2	0.07038	1.43 – 0.45/2 + 0.3
3.	Right hand side bed room						
	1st Footing						
	Long wall	1	3.95	0.65	0.4	1.027	3.3 + 65/2 + 0.65/2
	Short wall	2	2.58	0.65	0.4	0.6708 × 2 = 1.3416	3.23 – 0.65
	2nd Footing						
	Long wall	1	3.75	0.65	0.4	0.675	3.3 + 0.45
	Short wall	2	2.78	0.65	0.4	0.5004 × 2 = 1.0008	3.23 – 0.45
	Plinth wall above footing						
	Long wall	1	3.64	0.34	0.2	0.24752	3.3 + 0.34
	Short wall	2	2.89	0.34	0.2	0.19652 × 2 = 0.39304	3.23 – 0.34
4.	Kitchen						
	1st footing						
	Long wall	1	2.28	0.65	0.4	0.5928	2.43 – 0.45/2 + 0.65/2
		1	2.43	0.65	0.4	0.6318	2.43 – 0.65/2 + 0.65/2
	Short wall	1	1.95	0.65	0.4	0.507	2.6 – 0.65
	2nd Footing						
	Long wall	1	2.18	0.45	0.4	0.3424	2.43 – 0.95/2 + 0.4/2
		1	2.43	0.45	0.4	0.4374	2.43 – 0.45/2 + 0.4/2
	Short wall	1	2.15	0.45	0.4	0.387	2.6 – 0.45

Continued

Item No.	Particulars of Item	No.	Length	Breadth	Depth	Quantity	Explanation
	Plinth wall above Footing						
	Long wall	1	2.18	0.34	0.2	0.14824	2.43 – 0.95/2 + 0.4/2
	Short wall	1	2.48	0.34	0.2	0.16864	2.43 – 0.34/2 + 0.4/2
		1	2.205	0.34	0.2	0.14994	2.6 – 0.45/2 – 0.3/2
5.	R.H.S. of kitchen						
	1st footing						
	Short wall	1	2.515	0.65	0.4	0.6539	2.99 – 0.65/2 – 0.3/2
	2nd footing						
	Short wall	1	2.665	0.45	0.4	0.4797	2.99 – 0.45/2 – 0.2/2
	Plinth wall above footing						
	Short wall	1	2.745	0.34	0.2	0.18666	2.99 – 0.34/2 – 0.15
6.	Latrine						
	1st Footing						
	Long wall	1	1.98	0.65	0.4	0.518	1.33 + 0.65
		1	1.98	0.3	0.4	0.2376	1.33 + 0.65/2
	Short wall	2	0.895	0.65	0.4	0.2327 × 2 = 0.4654	1.37 – 0.65/2 – 0.3/2
	2nd Footing						
	Long wall	1	1.78	0.65	0.4	0.46	1.33 + 0.45
		1	1.78	0.2	0.4	0.1424	1.33 + 0.45
	Short wall	2	1.045	0.65	0.4	0.545	1.37 – 0.2/2 – 0.4/2
	Plinth wall above footing						
	Long wall	1	1.725	0.34	0.2	0.1173	1.33 + 0.45/2 + 0.3/2
		1	1.725	0.15	0.2	0.05175	1.33 + 0.45/2 + 0.3/2
	Short wall	2	1.07	0.34	0.2	0.07276 × 2 = 0.14552	1.37 – 0.15/2 – 0.4/2
7.	Bath						
	1st footing						
	Long wall	1	1.78	0.65	0.4	0.4628	2.43 – 0.65/2 – 0.8/2
		1	1.78	0.3	0.4	0.2136	2.43 – 0.65
	2nd footing						
	Long wall	1	1.98	0.45	0.4	0.3564	243 – 0.45
		1	1.98	0.2	0.4	0.1584	2.43 – 0.45
	Plinth wall above footing						
	Long wall	1	2.09	0.34	0.2	0.1421$\dot{2}$	2.43 – 0.34
		1	2.09	0.15	0.2	0.0627	2.43 – 0.34
					Total	23.962.3 m^3 24 m^3	
D.	2.5 cm Damp Proof Course						
1.	Drawing Room						
	Long wall	2	5.32	0.34	–	1.8088 × 2	plinth wall
		2	5.015	0.34	–	1.7051 × 2	
	Short wall	2	3.335	0.34	2.678	6.01	

Continued

Item No.	Particulars of Item	No.	Length	Breadth	Depth	Quantity	Explanation
2.	Front Verandah						
	Long wall	1	3.285	0.34	–	1.1169	
	Short wall	1	1.035	0.34	–	0.3519	
3.	Right hand side bed room						
	Long wall	1	3.64	0.34	–	1.2376	
	Short wall	2	2.89	0.34	–	0.9826 × 2 = 1.9652	
4.	Kitchen						
	Long wall	1	2.18	0.34	–	0.7412	
		1	2.48	0.34	–	0.8432	
	Short wall	1	2.205	0.34	–	0.7497	
5.	R. H. S. of kitchen						
	Short wall	1	2.745	0.34	–	0.9333	
6.	Latrine						
	Long wall	1	1.725	0.34	–	0.5865	
		1	1.725	0.15	–	0.25875	
	Short wall	2	1.07	0.34	–	0.7276	
7.	Bath						
	Long wall	1	2.09	0.34	–	0.7106	
		1	2.09	0.15	–	0.3135	
					Total =	18.05 m^2	
	Deduct						
	Door sill D_1	5	0.9	0.34	–	1.53	
	Door sill D_2	2	0.75	0.15	–	0.225	
		1	0.75	0.34	–	0.255	
	Total of deduction					2.010	
					Net Total =	16.840 m^2	
						say 17.0 m^2	
E.	1st Class Brick Work in Super Structure						
1.	Drawing Hall						
	Long wall	2	4.96	0.23	3.3	7.52928	4.73 + 0.23
	Short wall	2	3.50	0.23	3.3	5.313	3.73 – 0.23
2.	Right Hand Side Bed Rooms						
	Long Wall	1	4.96	0.23	3.3	3.76464	4.73 + 23
	Short wall	2	3.00	0.23	3.3	4.554	3.23 – 0.23
3.	Front verandah						
	Short wall	1	3.00	0.23	3.3	2.277	3.23 – 0.23
4.	Kitchen						
	Long wall	2	2.43	0.23	3.3	3.68874	2.43 + 0.23/2 – 0.23/2
	Short wall	1	2.37	0.23	3.3	0.179883	2.6 – 0.23

Continued

Item No.	Particulars of Item	No.	Length	Breadth	Depth	Quantity	Explanation
5.	Right hand side of kitchen						
	Short wall	1	2.82	0.23	3.3	2.14038	299 – 0.23/2 – 0.15/2
6.	Latrine						
	Short wall	2	1.54	0.23	3.3	2.33772	1.37 + 0.23/2 + 11/2
	Long wall	1	1.1	0.11	3.3	0.3993	1.33 – 0.23/2 – 0.23/2
		1	1.1	0.23	3.3	0.8349	
7.	Bath						
	Long wall	1	2.2	0.11	3.3	0.7986	2.43 – 0.23
		1	2.2	0.23	3.3	1.6698	
					Total =	37.10619 m^3	
	Deduct						
	Door opening D_1	5	2	0.9	0.23	2.07	
	Door opening D_2	3	2	0.75	0.23	1.035	
	Window opening W_1	2	0.23	0.20	1.2	0.1104	
	Window opening W_2	3	0.23	0.9	1.2	0.7452	
	Window opening W_3	1	0.23	0.6	1.2	0.1656	
	Ventilator opening V_1	2	0.23	2	0.5	0.46	
	Ventilator opening V_2	4	0.23	0.9	0.5	0.414	
	Ventilator opening V_3	1	0.23	0.6	0.5	0.069	
	Verandah opening	1	3.0	0.23	3.2	2.208	
					Grand total (for Brick work)	28.835 m^3	
						say ≃ 29 m^3	
	Lintels						
	Over Door D_1	5	1.2	0.23	0.12	0.165	0.9 + 0.3
	Over Door D_2	3	1.05	0.23	0.12	0.087	
	Over Window W_1	2	2.3	0.23	0.12	0.06348	2.0 + 0.3
	Over window W_2	3	1.2	0.23	0.12	0.099	
	Over Window W_3	1	0.9	0.23	0.12	0.02484	
					Net Total =	0.438 m^3	
F.	ROOFING IN R.C.C.						
	(Except latrine)	1	7.39	7.19	0.1	5.3134	
	Latrine	1	1.54	1.33	0.1	0.204	
					Total R.C.C.	5.517 m^2	
G.	FLOORING						
1.	Drawing Room						
	Stone work	1	4.5	3.5	0.1	1.575	

Continued

Item No.	Particulars of Item	No.	Length	Breadth	Depth	Quantity	Explanation
2.	Bed Room	1	3.07	3.0	0.1	0.921	
3.	Verandah	1	3.00	1.2	0.1	0.36	
4.	Kitchen	1	2.32	2.2	0.1	0.5104	
5.	Right hand side of kitchen	–	2.87	2.20	0.1	0.6314	
6.	Bath	1	2.20	1.2	0.1	0.204	
7.	Latrine	1	1.2	1.1	0.1	0.132	
					Total	4.3938 m^3	
						$\simeq$ 4.4 m^3	
	Brickwork in Parapet Wall						
	Long Walls	1	7.39	0.23	0.9	1.52973	
		1	8.72	0.23	0.9	1.80504	
	Short wall	1	6.96	0.23	0.9	1.44072	7.19 – 0.23
		1	6.96	0.23	0.9	1.44072	
		1	1.1	0.23	0.9	0.2277	
					Total =	6.44391 m^3	
						$\simeq$ 6.5 m^3	

Item No.	Párticular of Item	No.	Length	Width	Height	Vol/Area	Remark
H.	LIME CONCRETE IN ROOF 3″ (7.5 cm)		7.190	8.720	0.075	4.7 m^3	Cantilever projection of the roof on either sides, if any have been ignored.
G.	CEMENT PLASTER						
	Drawing Room						
	Long wall	2	4.50	–	3.3	29.7	
	Short wall	2	3.50	–	3.3	23.1	
	Verandah						
	Long wall	1	3.00	–	3.3	9.9	
	Short wall	2	1.43	–	3.3	9.45	
	Bed room						
	Long walls	2	3.07	–	3.3	20.3	
	Short walls	2	3.00	–	3.3	19.8	
	Kitchen						
	Long walls	2	2.32	–	3.3	15.3	
	Short walls	2	2.20	–	3.3	14.5	
	Lobby						
	Long wall	1	2.75	–	0.3	0.835	
	Short walls	2	2.70	–	3.3	17.8	

Continued

Item No.	Particular of Item	No.	Length	Width	Height	Vol/Area	Remark
	Bath						
	Long walls	2	2.70	–	3.3	17.8	
	Short walls	2	1.70	–	3.3	11.2	
	Latrine						
	Long walls	2	1.70	–	3.3	11.2	
	Short walls	2	1.10	–	3.3	7.26	
						212.0	
H.	WOOD WORK						
	Chaukhat for doors and windows						
	D_1 2 × 0.9 × 0.075 × 0.100 1.925 × 0.075 × 0.100	5				0.13875	Chaukhats 3″ × 4″
	D_2 2 × 0.75 × 0.075 × 0.100 1.925 × 0.075 × 0.100	3				0.0792	
	Windows						
	W_1 2 × 2.0 × 0.075 × 0.100 2 × 1.05 × 0.075 × 0.100	1				0.0457	
	W_2 2 × 1.05 × 0.075 × 0.0100 2 × 1.05 × 0.075 × 0.100	3				0.0457	
	W_3 2 × 0.6 × 0.075 × 0.100 2 × 1.05 × 0.075 × 0.100	1				0.0248	
	Clerestories (1000 × 500) V	5				0.37620	
	Long sides	2	1.0	0.075	0.100		
		2	0.35	0.075	0.100	0.10125	
						0.47745 m^3	
I.	SURFACE AREA OF TEAK PANEL						
	D_1	5	–	0.75	1.925	7.25	
	D_2	3	–	0.60	1.925	3.45	
	W_1	1	–	1.85	1.125	2.10	
	W_2	3	–	0.75	1.125	2.54	
	W_3	1	–	0.45	1.125	0.51	
						15.75 m^2	
	Glass Panes					3.15 m^2	Assuming 20% of the area glazed and provided with glass panes.

Abstract of Material and Cost

S. No.	Item	Units	Quantity	Rate	Cost	Analysis of Rates
1.	Earth Work of foundation	m³	42	Rs 8.94/m³	375.00	Rate = (0.358 A + 0.0358 B) Rs/m³, where 'A' and 'B' are rates of beldar and mason respectively, (A = Rs 20/- per day and Rs 40/- per day) Lime concrete Labour = 0.4 (A + B + C) A = Mason = 40/per day B = Labour = C = Bhisti = 20/- per day Labour = Rs 32/m³ (b) Material 1 m³ ballast = Rs 125/- 0.19 m³ Lime = Rs 50/- 0.19 m³ Surkhi = Rs 60/- 0.19 m³ Sand = Rs 25/- 260/- Total rate per m³ of lime concrete = 260.00 + 32.00 = 292.00 per m³
2.	Lime Concrete in foundation (1 : 1 : 1)	m³	6.26	Rs 292.00/m³	1830.00	
3.	Brick Work in foundation upto plinth (1 : 6 cement mortar)	m³	24.0	Rs 436.20/m³	10,450.00	
4.	D.P.C. (2.0 cm thick)	m²	12.0	Rs 24.50/m²	2940.00	
5.	I-Class brick work in super structure 1 : 4 cement mortar	m³	29	Rs 481.70/m³	13,900.00	
6.	R.C.C. Lintel	m³	0.183	Rs 1400.00/m³	265.00	
7.	R.C.C. in roof	m³	7.36	Rs 1400.00/m³	10,350.00	
8.*	Flooring - 2.5 cm laid in panels, overlaid on 8 cm thick base concrete	m³	4.4	Rs 54.50/m²	240.00	= 260.00 + 32.00 = 292 per m³
9.	Parapet wall Brick work in 1 : 4 cement mortar	m³	6.5	Rs 481.70/m³	33,050.00	
10.	Lime concrete on roof (1 : 1 : 1)	m³	4.7	Rs 376.80/m³	1,770.00	
11.	Cement Plaster 1 : 6 cement mortar, 0.25 cm thick	m²	183.35	Rs 25.73/m²	4,710.00	
12.	White washing one coat with 2 coats of distemper	m²	183.25	Rs 9.04 m²	1,660.00	
13.	Wood work					
	a. Chaukhat	m³	0.48	Rs 5,800/m³	2,780.00	
	b. Panels	1¾" thick teak	m²	15.99	Rs 300/m²	4,800.00
	c. Glass panes	m²	3.20	Rs 24.50/m²	78.50	
				Total	46544.50	

Additional Items

Rate Analysis

3. Brick work in foundation upto plinth

Brick work in 1 : 6 cement mortar

(a) materials

Cement bags 1.35 @ Rs 70/- per bag	=	94.50
Sand 0.275 m^3 @ Rs 125/- per m^3	=	22.70
Bricks 460 @ Rs 500/- per 1000	=	230.00
		347.20

(b) labour = (2.5 A + 5 B + 1/2 C + 1/4 D) × 0.4

A = Rate of Mason = Rs 40/- per day

B = Rate of Labour = Rs 20/- per day

C = Rate of Bhisti = Rs 20/- per day

D = Rate of Mistri = Rs 50/- per day

Labour = (100.00 + 100.00 + 10.00 + 12.50) × 0.4 = 222.50 × 0.4 = Rs 89.000

∴ Total = 347.20 + 89.00 = Rs 436.20/m^3

4. I class brickwork in super structure with 1 : 4 cement

Cement bags 2.00 @ 70/- per bag	=	Rs 140.00
Deduct cost of cement in item 3	=	94.50
Net		Rs 45.50

∴ Rate = 436.20 + 45.50 = Rs 481.70/m^3

5. R.C.C. (1 : 2 : 4) Rate/m^3

Material

(a) Cement - 5.76 bags @ Rs 70/- per bag	=	403.20
(b) Sand - 0.4 m^3 @ Rs 125/- per m^3	=	50.00
(c) coarse aggregate - 0.8 m^3 @ Rs 125/- per m^3	=	100.00
(d) Steel - 0.8 quintal @ 800/- per quintal	=	640.00
		1193.00

Labour

(i) Bending and cutting etc. (3/4 A + B) × 2.25 per quintal = (3/4 × 40 + 20) × 2.25 = 50 × 2.25 = 112.50

Labours for 0.8 quintal = Rs 90.000

(ii) Centering, shuttering and mixing etc.

= (2 A + 8 B + 2 C + 1/4 D) × 0.4 per 1.41 m^3

= (80 + 160 + 40 + 12.50) × 0.4 = 292.50 × 0.4 = Rs 117.00/m^3

Add 20% for electrification and sanitary provision	9,308.90
Add 5% for contingency	2,327.22
Total	58,180.62
	Say 58,200.00

Rate = Rs 58,200.00/m^3

Total 1193.00 + 90.00 + 117.00 = Rs 1400.00/m^3

6. Lime concrete on roof

a. Labour (A + 11 B + C + D/4) × 0.4/m^3

= (40 + 220 + 20 + 12) × 0.4

= 292 × 0.4 = Rs 116.80/m^3

b. Material = As above Rs 260.00

Total = Rs 376.80/m^3

7. Cement Plaster 1 : 6 cement mortar

1.25 cm thick cement

Plaster

a. Material - 0.079 cement bag @ Rs 70 per bag = 5.53

Sand - 0.016 m^3 @ Rs 120.00/- per m^3 = 2.00 (app.)

7.73

b. labour - (A + 2 B + 1/2 C) × 0.2

= (40 + 40 + 10) × 0.2

= 90 × 0.2 = 18.00

Total = Rs 25.73 per m^2

8. White-washing — one priming coat and 2 coats of dry distemper on new work

Rs 5.50 + 3.54 = Rs 9.04 per m^2

9. Wood Work

a. Chaukhats - Rs 5,800 per m^3

b. Fixing of CP teak panels 1¾″ thick - Rs 300.00 per m^2

c. Glass panes - Rs 24.50 per m^2

Example 5.8 Estimation of Six-roomed Building with Front Verandahs and Back Verandahs (Refer Fig. 5.8).

Details of Doors and Windows		
Item	Width (cm)	Height (cm)
D	120	200
D_1	110	200
D_2	80	180
W	100	150
W_1	180	150
W_2	75	100
C.W.	75	50
S	100	180

Abstract of Material Quantities

S. No.	Item	Qty	Units	Rate	Cost		Remark
					Rs	p.	
1.	Earth work in excavation	76.2	m^3	8.60/m^3			0.358 A + 0.0358 B = 7.16 + 1.432 = 8.59 Say 8.60/-
2.	Earth work in filling	103.0	m^3				
3.	Lime concrete in foundation	25.3	m^3	292.00/m^3			
4.	I class brick work in 1 : 6 mortar in foundation up to plinth	51.0	m^3	436.20/m^3			
5.	2.0 cm thick damp proof course	30.16	m^2	24.50/m^2			
6.	I class brick work in 1 : 4 cement mortar in super-structure	88.2	m^3	481.70/m^3			
7.	R.C.C. Work	31.7	m^3	1400.00			+900.00 M.T. for bending etc.
8.	7.5 cm lime concrete roof terracing	141.44	m^2	376.80/m^3			
9.	Bitumen layer of 2 coats on roofs	141.44	m^2	376.80/m^3			
10.	Cement plaster			25.73/m^2			
11.	White washing			9.09			
12.	Wood work						
	(a) Chowkhat			5800/m^3			
	(b) Panels			300/m^2			
	(c) Glass panes			24.30/m^2			

SIX-ROOMED BUILDING WITH FRONT AND BACK VERANDAH

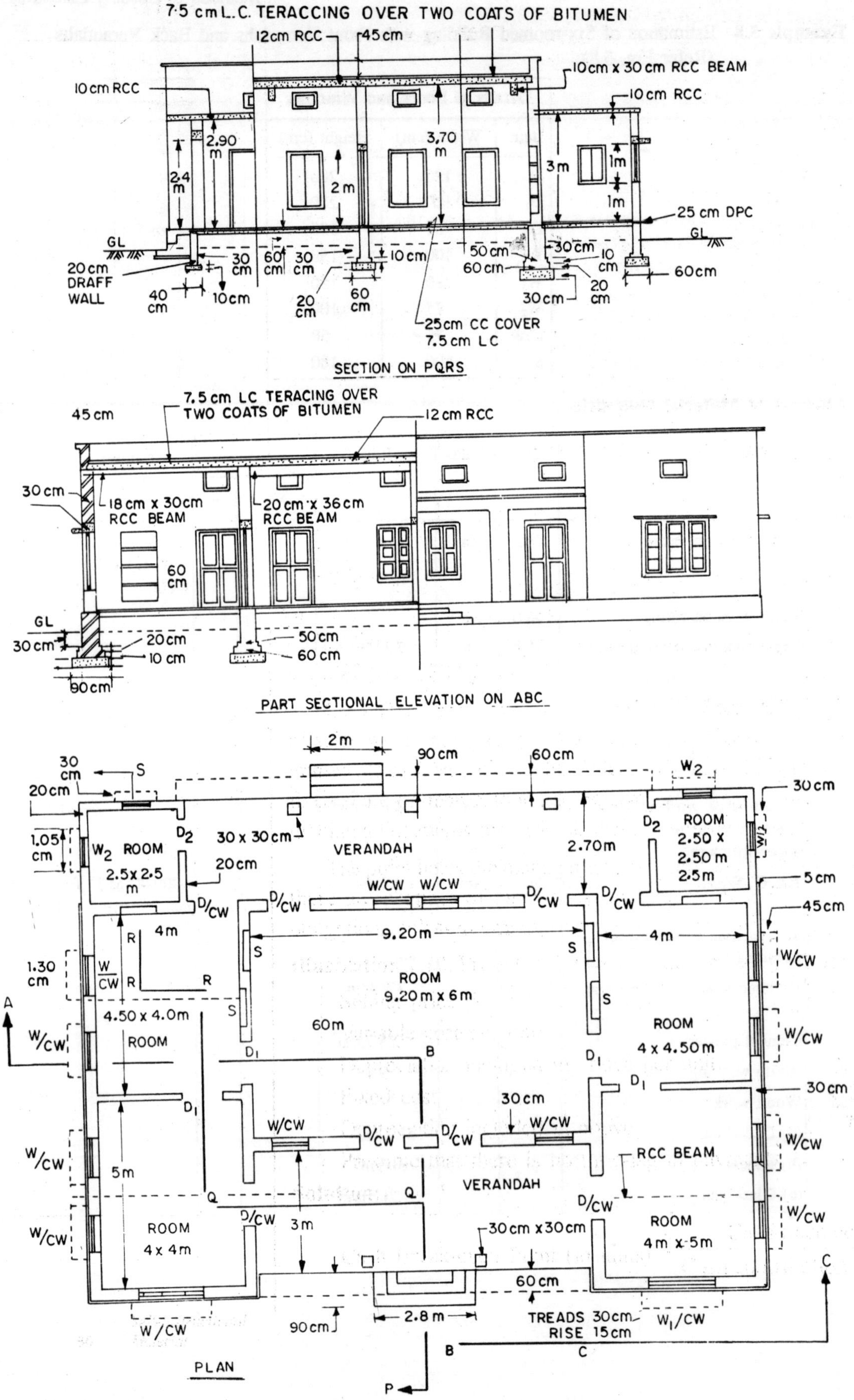

Fig. 5.8

Cement Plaster and White Washing

S. No.	Description	No.	L	B	H	Qty	Total
1.	Outer wall						
	(a) (5.0 + 0.3 + 4.5 + 0.20 + 0.3 + 2.5 + 0.20 = 13.0) side walls	2	13.00	3.7	–		
	(b) (2.5 + 0.2 + 0.2 + 2.5) 2 + 2 (4 – 2.5 – 0.2) + 0.3 + 9.2 + 0.3 – back wall = 5.4 × 2 + 2.6 + 9.8 = 10.8 + 2.6 + 9.8 = 23.2 m – backwall	1	23.2	3.7			
	(c) (4.0 + 0.3 + 0.3) × 2 + 2 × 3 + 9.2 = 9.2 + 6 + 9.2 = 24.4 m – front wall	1	24.4	3.7			
	(d) Rooms (Inner surface) (4.5 + 4 + 4.5 + 4) 2 + (2 × 4 + 2 × 4.5) 2 + (2 × 9.2 + 2 × 6) = 34 + 34 + 30.4 = 98.4 m	1	98.4	3.7			
	(e) short height walls = (2.5 × 4) = 10 m	1	10 m	3.0			

Centre to Centre Length of Walls.

Total centre to centre length of left side rooms and right side rooms.

$$= 2 \{2 (4.5 + 5.0 + 0.2 + 0.3) + 2 (4.0 + 0.30)\} = 57.20 \text{ m}$$

Total centre to centre length of central rooms

$$= 2 (4.5 + 4.5 + 0.2 + 0.3) = 2 \times 9.5 = 19.00 \text{ m}$$

Total centre to centre length of all 30 cm walls

$$= 57.20 + 19.00 = 76.20 \text{ m}$$

No of junction = 4 (with 30 cm wall)

= 20 cm. Wall

Total centre to centre length of 20 cm partition walls in rooms = 2 × 4.30 = 8.6 m

No of junction with 30 cm wall = 4

Total centre length of 20 cm wall back side room

$$= 2 \{2 \times 2.75 + 2.70\} = 16.40 \text{ m.}$$

No of junctions = 4 (with 30 cm wall)

Centre to centre length of front verandah = 9.50 meter

Centre to centre length of back verandah

$$= (4.0 + 4.5 + 4.5 + 4.0 + 4 \times 0.30 + 20) - (2 \times 2.90) + 0.20 = 12.80 \text{ m.}$$

Total centre length of parapets

$$= 2 \times 10 + 18.2 + 2 \times 4.5 + 9.40 + 2 \times 3.70 = 64.00 \text{ m.}$$

Item No.	Description of Item	No.	Length	Breadth	Height	Qnty.	Total
1.	Earth work in excavation						
	30 cm. walls of rooms	1	74.40	0.90	0.90	60.26	76.2 – 4 × 1/2 × 0.9 = 74.4
	20 cm. walls of back	1	6.8	0.60	0.90	3.672	8.6 – 4 × 0.45 = 6.8
	20 cm. walls of back	1	14.60	0.60	0.90	7.9	16.4 – 4 × 15 = 14.6
	Pillars						
	Front plinth dwarf wall	1	6.80	0.40	0.40	1.09	9.5 – 3 × 9 = 6.8
	Back (sum total length)	1	9.50	0.40	0.40	1.52	12.80 – 3 × 9 – 0.6 = 9.50
	Step front	1	2.90	0.95	0.15	0.41	5 cm projection
	Back step	1	2.10	0.95	0.15	0.30	
					Total	76.162 m^3	
						$\simeq$ 76.2 m^3	
2.	Earth work in filling						
	Left and right side rooms						
	(i)	2	3.90	4.45	0.525	18.22	
	(ii)	2	3.90	4.95	0.525	20.77	
	Central room	1	9.1	5.90	0.525	28.187	
	Back side room	2	2.500	2.45	0.525	6.43	
	Front verandah	1	9.10	2.75	0.525	13.14	
	Back verandah	1	12.60	2.45	0.525	16.21	
					Total	102.45 m^3	
						$\simeq$ 103.0 m^3	
3.	Lime concrete in foundation						
	30 cm. walls of rooms	1	74.40	0.90	0.30	20.09	
	20 cm wall of rooms	1	7.4	0.60	0.20	0.888	8.6 – 4 × 0.3 = 7.4
	20 cm walls of back small room	1	15.20	0.60	0.20	1.82	16.4 – 4 × 0.3 = 15.2
	Pillars						
	Plinth dwarf wall						
	Front (sum total length)	1	8.00	0.40	0.10	0.032	9.5 – 3 × 0.5 = 8.00
	Front step	1	2.90	0.95	0.10	0.28	
	Back step	1	2.10	0.95	0.10	0.20	
					Total	25.258 m^3	
						$\simeq$ 25.3 cum	

Continued

Item No.	Description of Item	No.	Length	Breadth	Height	Qnty.	Total
4.	1st Class brick work in 1 : 6 mortar in foundation upto plinth						
	30 cm. wall of rooms						
	1st footing	1	75.00	0.60	0.20	9.00	76.2 – 4 × 3 = 75.0
	2nd footing	1	75.20	0.50	0.10	3.76	76.2 – 4 × 0.2 = 75.20
	Plinth wall above footing	1	75.40	0.40	0.90	27.14	76.2 – 4 × = 75.40
	20 cm walls of rooms						
	Footing	1	7.60	0.30	0.10	0.228	8.6 – 4 × 0.25 = 7.6
	Plinth wall above footing	1	7.80	0.20	0.90	1.404	8.6 – 4 × 0.2 = 7.8
	20 cm walls of back small rooms	1	15.40	0.30	0.10	0.46	16.4 – 4 × 15.4
	Plinth wall above footing	1	15.60	0.20	0.90	2.81	16.4 – 4 × 2 = 15.60 m
	Pillars						
	1st footing	5	0.6	0.6	0.20	0.36	
	2nd footing	5	0.5	0.5	0.10	0.13	
	Plinth wall above footing	5	0.4	0.4	0.90	0.72	
	Verandah plinth dwarf wall -						
	Front (sum total length)	1	8.30	0.20	0.90	1.49	9.5 – 3 × 0.4 = 8.30
	Back (sum total length)	1	11.40	0.20	0.90	2.05	12.8 – 3 × 9 – 0.2 = 11.4
	Step front						
	1st step	1	2.80	0.90	0.15	0.38	
	2nd step	1	2.20	0.60	0.15	0.20	
	3rd step	1	1.66	0.30	0.15	0.07	
	Step back						
	1st step	1	2.00	0.90	0.15	0.27	
	2nd step	1	2.00	0.60	0.15	0.18	
	3rd step	1	2.00	0.30	0.15	0.09	
					Total	50.742 cum	
						≃ 51.0 cum	
5.	2.5 cm damp proof course						
	30 cm. walls of room	1	75.40	0.40	–	30.16	
	20 cm. of walls of rooms	1	7.8	0.20	–	1.56	
	20 cm. walls of back rooms	1	15.60	0.20	–	3.12	

Continued

Item No	Description of Item	No.	Length	Breadth	Height	Qnty.	Total
	Pillars	5	0.40	0.40	–	0.80	
					Total	35.64 cum	
	Deduct door sills.						
	D in 30 cm wall	8	1.20	0.40	–	3.84	
	D_1 in 30 cm wall	2	1.10	0.40	–	0.88	
	D_1 in 20 cm wall	2	1.10	0.20	–	0.44	
	D_2 in 20 cm wall	2	0.80	0.20	–	0.32	
					Total =	5.48 Sq.m.	
					Net total	30.16 Sq.m.	
6.	1st Class brick work in lime mortar in super structure						
	30 cm walls of rooms	1	75.60	0.30	3.70	83.92	76.2 – 4 × 3 = 75.60 m
	20 cm walls of rooms	1	8.00	0.20	3.70	5.92	8.6 – 4 × 0.15 = 8.0
	20 cm walls of back rooms	1	15.80	0.20	2.90	9.16	16.4 – 4 × 0.15 = 15.80 cm
	Pillars	5	0.30	0.30	2.40	1.08	
	Verandah 30 cm wall above pillars above lintel						
	Front	1	9.20	0.30	0.30	0.83	9.5 – 3 = 9.20
	Back	1	12.50	0.30	0.30	1.13	12.8 – 0.3 = 12.50
	Parapet 20 cm wall	1	64.00	0.20	0.545	6.98	Ht = 0.35 T = 0.075
	Upper 30 cm coping	1	64.00	0.30	0.10	1.92	+ 1.2 = 0.545
					Total	110.90 Cu.m. ≃ 111.0 Cu.m.	
	Deduct						
	Door openings						
	D in 30 cm wall	8	1.20	0.30	2.00	5.76	
	D_1 in 30 cm wall	2	1.10	0.30	2.00	1.32	
	D_1 in 20 cm wall	2	1.10	0.20	2.00	0.88	
	D_2 in 20 cm wall	2	0.80	0.20	1.80	0.58	
	Window openings						
	W in 30 cm wall	12	1.00	0.30	1.5	5.40	
	W_1 in 30 cm wall	0.2	1.80	0.30	1.5	1.62	
	W_2 in 20 cm wall	4	0.75	0.20	1.0	0.60	
	C.W. opening in 30 cm wall	2.2	0.75	0.30	0.5	2.48	
	Shelve openings	6	1.00	0.20	1.80	2.16	Back wall 10 cm thick

Continued

Item No.	Description of Item	No.	Length	Breadth	Height	Qnty.	Total
	R.C.C. lintels						
	Over doors - D	3	1.40	0.30	0.10	0.126	10 cm is the bearing
	Over doors - D_1	2	1.30	0.30	0.10	0.078	
	Over doors - D_1	2	1.30	0.20	0.10	0.052	
	Over doors - D_2	2	1.00	0.21	0.10	0.040	
	Over windows - W	12	1.20	0.30	0.10	0.432	
	Over windows - W_1	2	2.00	0.30	0.10	0.120	
	Over windows - W_2	4	0.95	0.20	0.10	0.076	
	Over C.S. windows	22	0.95	0.20	0.10	0.418	
	Over shelves	6	1.20	0.30	0.10	0.216	
	Bearings of verandah lintels	2 × 2	0.15	0.30	0.20	0.036	
					Total	2.813 Cu.m.	
					Net Total	88.127 Cu.m.	
					≃	88.2 Cu.m.	
7.	R.C.C. Work excluding steel and bending but with centering, shuttering and binding steel						
	R.C.C. Beam	4	4.40	0.18	0.30	0.950	20 cm bearing
		1	9.6	0.20	0.36	0.691	
	R.C.C. Roof slab						
	Left side and right side rooms	2	10.00	4.30	0.12	10.320	15 cm bearing
	Central rooms	1	9.50	6.30	0.12	7.182	including cm projection
	Back side rooms	2	2.95	2.95	0.10	1.741	
	Verandah						
	Front	1	9.50	2.95	0.10	2.803	excluding chhujja
	Back	1	12.80	2.65	0.10	3.392	
	Chhujja front	1	9.50	0.65	0.06	0.371	6 cm thickness
	Chhujja back	1	13.00	0.65	0.06	0.507	
	Sun shades						
	Over window W	8	1.30	0.45	0.05	0.234	15 cm extended on both ends
	Over window W_1	2	2.10	0.45	0.05	0.095	
	Over window W_2	4	1.05	0.80	0.05	0.063	
					Total	28.349 Cu.m.	
	Lintels						
	Over front verandah	1	9.50	0.30	0.20	0.570	15 cm bearing
	Over back verandah pillars	1	13.00	0.30	0.20	0.780	15 cm bearing

Continued

Item No.	Description of Item	No.	Length	Breadth	Height	Qnty.	Total
	Over doors and windows	as calculated above in item 6				2.003	
					Total	31.702 Cu.m.	
8.	Mild steel work including bending in reinforcement						
	1% of R.C.C. work	$\frac{31.702}{100} \times 1$				= 0.31702 Cu.m.	
						≃ 0.32 Cu.m.	
9.	7.5 cm Lime Concrete roof terracing						
	Left and right hand side rooms	2	9.90	4.20	–	83.16	10 cm
	Central rooms	1	9.60	6.20	–	58.28	
					Total	141.44 Sq.m.	
10.	Bitumen insulation layer of two coats roof						
	Left and right hand side rooms	2	9.90	4.20		83.16	
	Central rooms	1	9.60	6.20		58.28	
					Total	141.44 sq. m.	

6

VALUATION

The main purpose of the process of valuation is to arrive at a fair and reasonable value of any existing engineering structure at present. In modern times, the relative values of various commodities are exposed in terms of money. Valuation is required for the following :

(i) Buying and selling of properties;
(ii) Taxation;
(iii) Security of loans;
(iv) Rent determination;
(v) Speculation;
(vi) Private development;
(vii) Betterment charges;
(viii) Insurance;
(ix) Reinstatement.

Classification of Property

A property is classified in the following two broad categories :

(i) Land which is an immovable property;
(ii) A property other than land is known as personal property.

As far as the valuer is concerned, the interests in land or land and buildings can be divided into two categories. (a) Freehold; (b) Lease hold. These terms are self-explanatory.

The principal types of lease are as follows :

(a) *Building Lease :* If the owner of a freehold plot of land commanding a good site for the construction of building is unable to develop his land, he gives the plot of land on lease to somebody who is eager to construct a suitable building at the site. It is clear that the lease holder will have to invest a good amount in the construction of the building over the plot of land and hence such lease will not become popular, if sufficient time is not given to the lease holder to enjoy the fruits of his investment. Building lease, therefore, is normally granted for 99 or 999 years.

(b) *Occupation Lease :* In this type of lease, the owner constructs a suitable structure on his land and together with land, he grants lease for certain occupation such as residential, factory, shop, office, etc. Following points are to be noted in case of the occupation lease.

(i) *Period of Lease :* The period of the occupation lease will depend on the use of the structure in his hand. In case of residential buildings, the lease period is about 2 to 5 years or so.

If the rent agreed in an occupation lease is very nearly equal to the prevailing rent for such similar land and structure, it is known as rack rent. But if the agreed rent is less than the prevailing rent, for such similar land and structure, it is known as head rent.

(ii) *Rising Rents :* Sometimes, in order to take the advantage of future development of the nearby areas, the owner grants the occupation lease on a rising rent basis.

(iii) Depending on the terms and conditions mentioned in the original lease, the lease holder may grant sub-lease to other persons.

(iv) *Life Lease :* In this type of the lease, the lease is granted for the duration of the life of the lease holder or some other person.

(v) *Perpetual Lease :* This type of lease is for a term of years. But it is renewable from time to time.

Sinking Fund : At the expiry of the period of utility, a property is either to be replaced or reconstructed. The fund set aside for this purpose, is known as sinking fund.

The amount of the annual instalment of the sinking fund may be found out from the following formula :

$$I = \frac{S\,i}{(1+i)^n - 1}$$

where S = amount of sinking fund

i = rate of interest

n = number of years required

I = annual instalment period.

Amortization : When the payment of a debt is made by a series of equal periodic payments, it is known as amortization. Each periodic payment includes two components

(i) Certain portion of the principal

(ii) An amount of interest on the outstanding debt

Scrap Value and Salvage Value : At the end of the period of usefulness, a property is discarded and then it can be scrapped or broken up into suitable units for disposal. The amount obtained by selling such units is known as scrap value.

Sometimes the property after being discarded at utility period, is sold as it is without being broken into pieces, and the amount realised over and above the cost of its removal and sale is known as the salvage value of the property.

Normally the scrap value or the salvage value of an asset or property has got some positive figure. But it may be zero or negative in some cases.

Methods of Valuation

For finding out the value of a particular property, the usual procedure to be followed is as follows.

(i) The net income after deducting all outgoings is found out.

(ii) A suitable rate of interest prevailing in the market for each type of property is assumed.

(iii) Based on the above rate of interest, the suitable figure for the Y.P. is obtained.

(iv) The net income, multiplied by Y.P. gives the capitalised value of the property.

This method is known as rental method of valuation and the result obtained by this method is the value of land and buildings, taken together for perpetual income.

In some cases, however, it becomes practically impossible to follow the above procedure. There may be non-availability of the rental value or the appropriate figure of rate of interest or there may be no transaction over a long period or the property may be meant for certain trades only. In such cases, the following methods are adopted :

(i) Direct comparisons of the capital value.
(ii) Valuation by reference to profits.
(iii) Valuation based on the cost.
(iv) Residual or Development Method.

Rent Fixation : Sometimes, it is required to fix up the rent of a property. The capitalized value of the property is known and dividing this capitalized value by a proper figure of year's purchase, the net income from the property can be worked out. All possible outgoings are added to this net income which will then give the gross rent or gross income per year from the property. Dividing this gross income by 12, rent per month can be calculated. The rent worked out by this procedure is known as a standard rent, while the actual rent of the property may be higher or lower than this rent, depending upon various uncertain elements, such as situation of the property, type of construction, demand and supply of such properties etc.

Market Value and Book Value : Market value of a property is defined as an amount which can be reasonably realised at any time from the open market when the property is put in the market for sale.

Book Value : Book value is defined as an amount shown in the account book of a particular year after making due provisions for the depreciation of the previous years.

Depreciation and Obsolescence : A property gradually loses some value due to its wear, tear and similar other causes. This loss in the value of a property is defined as depreciation. Methods adopted to calculate the depreciation of a property are as following :

(i) *Straight Time Method :* In this method it is assumed that a property loses its value by the same amount every year.

Let n = life of the property in years
C = original cost
S = salvage or scrap value
D = annual depreciation.

Then, $$D = \frac{C - S}{n}$$

(ii) *Constant Percentage Method :* In this method it is assumed that a property will loose its value by a constant percentage of its value at the beginning of every year. Let n, C and S represent as above, and let p = constant percentage, then

value of property at the end of the first year $= C - p \cdot C = C\,(1 - p)$
value of property at the end of the second year $= C\,(1-p) - p \cdot C\,(1-p) = C\,(1-p)^2$
value of property after 2 years $= C\,(1-p)^2$
Similarly, the value of property after n years $= C\,(1-p)^n$

$$\therefore \quad C\,(1-p)^n = S$$

or $$p = 1 - \left(\frac{S}{C}\right)^{1/n}$$

(iii) *Unit Cost Method :* This method of calculating depreciation is applicable to machines and other equipments where it is possible to work out the cost of unit output. In this method it is assumed that a machine after n years of use shall give the same unit cost of output during its remaining years of life.

(iv) *Quantity Survey Method :* This is the most widely used method in practice. The property is studied in detail and loss in value owing to physical deterioration is worked out. Each and every step is based on some logical ground without any relation to some fixed percentage to the cost of the property. Thus the basis of this method is observation and not assumption.

Sinking Fund Method : In this method the depreciation of the property is assumed to be equal to the annual sinking fund plus the interest on the fund for that year.

Outgoings : Practically all types of the property are subject to various kinds of outgoings which are expenses to be incurred to maintain the revenue from the property. It is very important for the valuer to estimate the outgoings as correctly as possible because on it depends the net income of property.

Usual types of the outgoings are :

(a) Municipal or local taxes;
(b) Government taxes;
(c) Repairs;
(d) Sinking fund;
(e) Management charges;
(f) Insurance charges;
(g) Loss of rent;
(h) Miscellaneous charges.

Gross Income and Net Income : The total amount of revenue received from a property during a year is known as gross income.

The net income from a property represents an amount left at the end of the year after deducting all usual outgoings.

Capitalised Value : It is defined as the amount of money whose annual interest at the highest prevailing rate of interest will be equal to the net income arising from the property.

Year's Purchase : Year's purchase is defined as the figure or multiplier, which when multiplied by the net income gives the capitalised value of a property at present time.

Capitalised value of a property = year's purchase × net income from the property.

$$\text{Year's purchase} = \frac{100}{\text{rate of interest}}$$

or

$$\text{Y.P.} = \frac{1}{i}$$

Single Rate and Dual Rate for Y.P. : In case of income for short duration, we have to take into consideration two rates of interest, while fixing the value for Y.P. :

(i) rate of interest for capital; and
(ii) rate of interest for redemption of capital.

Now, if these two rates are same, it is said that Y.P. is based on single rate basis.

7

SPECIFICATIONS OF IMPORTANT ITEMS

The specifications of the work are specified by the Engineer-in-charge before the construction work is taken up. The specifications are prepared by the Engineer and form essential part of the agreement. While preparing specifications of the different items of the work, the Engineer-in-charge has to keep the following in mind:

(i) *Subject Matter:* The subject matter should specify the information required after the contract is awarded. Use of cross reference should be reduced to a minimum. Thus every specification should be complete in itself.

While writing specifications only suitable words with desired meaning should be used. The informations given in specifications should be complete and correct avoiding ambiguity altogether. Also, care should be taken to see that the information is not repeated. The specifications should clearly lay down the following:

(a) Subject matter.
(b) Ensure proper words.
(c) Accuracy as desired.
(d) Practical limits and commercial sizes.
(e) The specifications should be clear, avoiding ambiguity, etc.
(f) The specifications should be fair and their achievement should be practically possible in the execution of the work.
(g) Brevity — The sentences of the specifications should be short, simple and concise.

Types of Specifications

The specifications are broadly divided into two categories:

(i) Brief specifications; and (ii) detailed specifications.

While preparing the estimate, the quantity surveyor writes down the brief description of every item. These are known as brief specifications and they may not form part of the contract document. They are useful only to the supervisory staff.

The detailed specifications form a part of the contract document and these are divided into three groups:

(i) General provisions;
(ii) Technical provisions;
(iii) Standard specifications.

(i) *General Provisions :* These are virtually the conditions of the contract and they apply to the work as a whole. In this document the conditions, governing the contract are written.

(ii) *Technical Provisions :* These specifications describe the technical requirement of each type of construction. The technical provisions are of 3 types :

(a) Specifications for material and workmanship;
(b) Specifications for performance;
(c) Specifications for proprietary commodities.

(a) *Specifications for Material and Workmanship :* These specifications are written for engineering structures such as bridges, buildings, etc., where it is not possible to conduct any practical test.

For materials, the following properties should be included in the specifications :

(i) Physical properties;
(ii) Chemical composition;
(iii) Electrical, thermal and acoustical properties;
(iv) Appearance;
(v) Various tests required for the material;
(vi) Information regarding protection required during transport.

For workmanship, the following important features should be included :

(i) The results desired should be specified, while keeping in view the practical limits of tolerance.
(ii) Detailed description of the construction method for each item should be written.
(iii) Instructions regarding the protection of the finished work should also be specified.

Specifications are seldom written completely for all items of work, but are left to the skill and competence of the supervisory staff.

Specifications of some of the important engineering works are given as below, only for the guidance of the readers.

Specification for Excavation

The sides of the foundation trenches shall be truly vertical and bottom shall be uniformly levelled. The excavated material shall be stacked away from the sides of the trenches of the excavation by at least 2 m. The excavated material shall be filled in the plinth in layers of 30 cm and well watered. The surplus excavated material shall be spread out uniformly upto a lead of 100 m. All the existing services such as water pipes, sewers, electric cables etc. which are met with in the foundation trenches, shall be carefully supported and protected by the contractor as per instructions of the Engineer-in-charge.

The excavation shall be measured as per exact length and width of the lowest step of the footings according to drawing or the Engineer's instructions. The depth of excavation shall be measured vertically from the average ground level, taken at site before starting of the work. The rate of excavation shall include all timber and other supports, which may be necessary for securing the sides of the trenches and the contractor shall be responsible for their safety. No extra amount shall be payable to the contractor for the excavation of the working space. The materials or values found during excavation shall be the property of the owner.

Specification for removal of water from foundation

All the foundation trenches should be clear of water. All the water getting admission to the trenches shall be pumped out so as to keep the trenches dry. The sumps shall be arranged such that water clear of sand is pumped out.

Brickbat lime concrete for foundation (1 : 2 : 4)

The proportion of brickbat lime concrete, shall be 1 part of lime mortar (1 : 2) and two parts of brickbat by volume. The lime mortar shall be prepared from 1 part of slaked lime and two parts of sand by volume. Lime to be used shall not give a residue of more than 10% by weight when treated with HCl. Sand to be used, shall be clean and coarse and shall not contain any organic impurities. Lime and sand shall be thoroughly mixed in a bullock driven mortar mixer fitted with Beale's tell tale and situated at site of work. Lime mortar shall be prepared in required quantities and shall be stacked under a shed and shall be kept damp at all times. Lime mortar which has become stale shall not be used for the work.

Brickbats

Brickbats shall be obtained from thorough burnt new bricks and shall be broken in sizes of 40 mm to 50 mm. Brickbats shall be free from dirt, dust or such other impurities. Mixing shall be done on a water tight platform. The materials, shall be turned over twice dry and twice again with water, added in required quantities so that whole surface of broken brickbats gets coated with mortar.

Placing of concrete shall start after the bottom of foundations is well dressed, watered and rammed by means of rammers. The concrete shall be laid in layers of 70 cm thickness and each layer shall be well rammed and well watered before next layer is laid. Concrete shall not be allowed to be thrown from top but shall be gently lowered in the trenches. Ramming shall be done till concrete becomes solid and compact. Watering of concrete surface shall be done at least for one week.

Brick work I class in cum. (1 : 6)

Bricks shall be table moulded, well burnt in approved kiln, copper coloured, free from cracks and with sharp and square edges. Bricks shall be uniform in shape and shall be of standard size and shall give clear ringing sound, when struck with each other. Bricks shall be well soaked in water for at least 12 hrs, before their use, preferably in a tank provided at site of work. The proportion of mortar shall be one part of cement to six parts of sand by volume. Cement shall be of standard quality. Sand shall be clean and coarse. Cement mortar shall be prepared in required quantity of not more than one bag of cement at a time and the quantity shall be consumed in half an hour after mixing. The mixing of mortar shall be done on a water tight platform or a steel trough. Water fit for drinking shall used in preparing mortar.

Broken bricks shall not be used except as closers. All corners shall be truly in plumb and sides shall be in line and level. Mortar joint shall break for bonding and shall not exceed 10 mm in thickness. All brick work shall be carried out in such a way that no portion is raised unduly above another. The length and height shall be measured as on at site. The thickness of walls shall be paid for as one brick, one and a half bricks, two bricks and so on. The rate for brick work includes necessary scaffolding also.

Timber doors and windows

Timber to be used shall be of best quality, well seasoned and free from cracks, knots, flares shakes and other defects. The rates for doors and windows includes fabricating and fixing in position of doors and windows. The shutters shall be prepared from scantlings 40 mm in thickness as per design supplied. The shutters shall be fully panelled and partly glazed as per plan supplied. The thickness of panels shall be 20 mm.

The frames for doors and windows shall be of 10 cm × 80 cm size and they shall be provided with M.S. hold fasts of section 30 mm × 6 mm and length 20 cm. Three number of hold fasts shall be provided on each side of the door frame and two such numbers of hold fasts shall be provided on each side of the door frame and two such number of hold fasts shall be provided on each side of the window frame. The sides of the frame abutting to the brick wall shall be coated with chocolate paint or coaltar. The workmanship of timber shall be of best quality and only skilled carpenters shall be employed on the work. If any undue shrinkage or bad workmanship is found out afterwards, the same shall be repaired by the contractor at his own cost.

The rate for timber work includes all the required fastenings and fixture which shall be of brass and of approved quality, design and make and they shall be fixed as directed. The rate for timber work includes 3 coats of oil paint of approved quality. The measurements of the work shall be taken overall including frames and no extra amount shall be paid for any wastage of material.

Specification for R.C.C. work (1 : 2 : 4)

The proportion of cement concrete shall be of one part cement two parts of sand and four parts of aggregate by weight. The cement to be used in this work shall comply with the standard requirements. The sand to be taken shall be clean and coarse and shall be free from any organic or vegetable matter. The sand shall be washed, if there is any trace of earth in it. Water to be used in concrete work shall be clean and fresh.

All reinforcement shall be of steel which shall comply with the standard requirements. All bars shall be placed as per design given by the Engineer and utmost care shall be taken to keep them in same position while packing concrete around them. Before laying the concrete, the reinforcement shall be got approved by the Engineer.

The centering for the concrete work shall be efficiently strong and rigid and in good condition so as to turn out a good smooth surface. The centering shall be removed only after obtaining the written permission of the Engineer.

The mixing of concrete shall be done in a mechanical mixer or by hand operation depending on the quantity of the concrete which shall be decided by the Engineer.

Before starting laying of concrete in position, the centering shall be well-watered. Concrete shall be laid in forms and tapped in such a way that no honey comb surface appears on removal of the forms. All edges and corners of the concrete surface shall remain unbroken, sharp and straight in line. The cement concrete shall be kept well watered for at least 20 days.

All exposed surfaces of the concrete work shall be finished with 12 mm thick cement plaster of sagol finish and three coats of white or colour wash shall be applied on these exposed surfaces of concrete. The items of plaster and white or colour wash shall be carried out as per specifications.

Pointing in cement 1 : 1

The pointing work shall be started as soon as the brick work is completed. The joints shall be racked out to a depth of 20 mm and the surface of the wall shall be cleaned, washed and well watered at least for 2 days.

20 mm Sand faced cement plaster

The work of sand faced cement plaster shall be carried out after masonry joints are racked out to a depth of 20 mm and the walls are well watered.

The first coat for sand faced cement plaster shall be of cement mortar of proportion 1 : 4. Mortar of 1 cement bag shall only be prepared at a time and it shall be prepared on a water tight platform or in a steel trough. The mortar shall be consumed in 30 mts after adding water. The first coat of plaster shall be uniformly applied in the best workmanship manner after watering the surface of the walls in advance and putting dots, and preparing sereeds for maintaining uniform thickness of plastered surface. The thickness of first coat shall be nowhere less than 12 mm. The first coat shall be well watered at least for 7 days. The first coat shall be provided with zigzag lines such that the first coat adheres well with the second coat.

The second coat of sand faced cement plaster shall be carried out in cement mortar of proportion 1 : 1 and shall be of thickness 8 mm. The second coat shall be applied on the first coat at least after 7 days of completion of the first coat. The second coat shall give uniform grey colour to the whole structure or any other colour as approved by the Engineer-in-charge. Sand to be used in the mortar for the second coat shall be perfectly screened so that the sand of uniform size appears on the surface. Sponge shall be used in the second coat and it shall be so worked that the density of sand grains appearing on the surface shall be equal and uniform. After completion of the second coat the surface shall be kept well watered at least for 15 days.

Specifications for white wash

Walls to be white washed shall be cleaned before the work is started. Fresh white lime shall be slaked at site of work and shall be mixed thoroughly with sufficient quantity of water in a tub and shall then be screened through a clean cloth. Clean gum or rice dissolved in hot water shall be mixed with lime in required quantities. The white wash shall be applied with jute brush and the brush shall be so worked that a surface with uniform colour is obtained. Three coats shall be applied, each after the previous coat has completely dried.

The stains of white wash on the floor or other surfaces shall be removed.

Fixing asbestos cement corrugated roofing sheets on steel purlins

The asbestos cement sheets to be used shall be of best quality and free from cracks. The sheets shall be laid with a side lap of at least one corrugation with a minimum of 15 cm end lap.

Fixing of sheets with steel purlins shall be done by 8 mm diameter galvanized hook-bolts. All hook bolts shall be provided with curved diamond washers. Expansion joints shall be provided on long stretches of roof. All holes for fixing sheets shall always be drilled and not punched. The payment shall be made for actual area covered by the asbestos sheets and no amount shall be given for wastage, laps, corrugation etc. The payment of ridges, hips and valleys shall be on running length.

Specifications for 16 cm water-bound macadam road for heavy traffic

The road metal shall be brought from approved quarry. In direct composition test, the road metal shall give a strength of 1450 kg/cm^2. The abrasion number of road metal shall not be less than 17 and the attrition number shall be between 7 and 8. In impact test, the road metal shall withstand a height of 18 to 20 cm and in cementation test, it shall withstand 100 blows.

The sub-grade shall be prepared to the camber of 1 : 36, the sub-grade shall be well dressed and a base coat of sand of depth 10 cm shall be placed. The road metal shall be placed on the finished sub-grade and properly compacted in two layers. The thickness of each layer shall be 12 cm, so that after consolidation, the thickness shall become 16 cm.

Each layer of the road metal shall be rolled dry so that a dense and properly inter-locked surface is obtained. Rolling shall be continued till the stones do not move under the weight of the roller. The roller shall pass equally over the entire surface and its speed shall be slow and uniform. The roller shall start from the edges of the road and shall be taken to the crown. Each successive strip shall overlap the proceeding strip to avoid formations of the weak points.

The finishing coat shall consist of sand blindage and then the required amount of water shall be spread over the surface. The rolling shall then be started and continued till a loaded cart goes over the finished surface without making any mark. The road shall then be opened for traffic.

Specification for Brick work II class in 1 : 6 cement mortar

Properties of bricks which are to be ground moulded and burnt in clamps — soaking in water — mortar — workmanship, mortar joints not to exceed 12 mm in thickness.

Specification for reinforced brickwork for wall in cement mortar (1 : 3)

Properties of bricks which are to be table moulded and burnt in kilns — soaking in water — mortar — reinforcement of 6 mm steel bars at the rate of one bar for half-brick thickness of wall and to be placed at every third coarse, mortar joints should not exceed 10 mm in thickness and reinforcement to be given a cover of 12 mm in all directions.

Specifications for Coarsed Rubble Masonry

Properties of stones to be of equal height, and face stones to be dressed by hammer — soaking in water — mortar — workmanship, each course to be of equal height and thickness of the joints not to exceed 10 mm.

Specification for 12 mm thick cement plaster in cement mortar (1 : 4)

Preliminary process of first coat — second coat of sagol finish after 6 hours of first coat — watering — three coats of white or colour wash.

Specifications for mosaic tiles

Properties of tiles such as shape, type, make, etc. — Cement mortar (1 : 3) — preparation of bedding with 12 cm thick B.B.L.C. of proportion 1 : 2 : 4 and to mention that the rate includes this bedding — workmanship — polishing — measurement.

Specifications for 15 cm thick cement concrete road 1 : 2 : 4 laid in single layer

Proportion — Cement — Fine aggregate — coarse aggregate — reinforcement (if any) — preparation of sub-grade-layer of sand on sub-grade, depth about 50 to 70 mm and to be rammed and well watered — centering — mixing — laying provision of expansion and contraction joints — curing cleaning measurement.

Specifications for cement Macadam Road on existing W.B. road surface

Materials, coarse aggregate (size 57% of 50 mm to 25 mm, and 25% of 12 mm to 6 mm), sand, cement and water, dry cement mortar proportion 1 : 2 — preparation of the road surface — placing of a thin layer of sand 12 mm thick on the sub-grade — placing of the forms — placing of the first layer of stone, depth 57 mm to 64 mm and rolling — spreading of the dry mortar, depth 57 mm to 64 mm application of water at the rate of 20 to 22 litres per sq.m — rolling — finishing — curing — measurement.

Specification for fabricating steel roof truss by welding

Members to be of section conforming to design — properties of steel to be used — gusset plate — process of welding — one coat of red lead — measurement to be on the basis of the weight of the truss.